NUCLEAR INSTRUMENTATION DEPARTMENT

ELECTRONICS AND WAVES—*A series of monographs*
EDITOR: D. W. Fry (Harwell)

SIGNAL, NOISE AND RESOLUTION
IN NUCLEAR COUNTER
AMPLIFIERS

SIGNAL, NOISE AND RESOLUTION IN NUCLEAR COUNTER AMPLIFIERS

By

A. B. GILLESPIE, B.Sc. (Eng.)

Atomic Energy Research Establishment, Harwell

NEW YORK: McGRAW-HILL BOOK CO., INC.
LONDON: PERGAMON PRESS LTD.
1953

Published in Great Britain by Pergamon Press Ltd., 2, 3 & 5 Studio Place, London, S.W.1
Published in U.S.A. by McGraw-Hill Book Co., Inc., 330 West 42nd Street, New York 36, N.Y.

Made in Great Britain at the Pitman Press

CONTENTS

	PAGE
EDITOR'S PREFACE	vii
AUTHOR'S PREFACE	vii
LIST OF SYMBOLS	ix

1 INTRODUCTION 1

2 NATURE OF THE SIGNALS FROM IONISATION
  CHAMBERS 4
  2.1 Ion Chambers 4
  2.2 Electron Chambers 10
  2.3 Effect of Amplifier Time Constants on Pulse
      Amplitude and Pulse Width 14

3 VALVE AND CIRCUIT NOISE 19
  3.1 General 19
  3.2 Thermal Noise 21
  3.3 Shot Noise 24
  3.4 Grid Current Noise 29
  3.5 Flicker Noise 40
  3.6 Other Forms of Noise 44
      3.6.1. Induced Grid Noise 44
      3.6.2 Thermal Noise in Anode Load of First Valve 46
      3.6.3 Current Noise in Semi-conductors 47
      3.6.4 Current Noise in Anode Load of First Valve 48

4 SIGNAL TO NOISE RATIO 49
  4.1 General 49
  4.2 Signal to Noise Ratio in Triodes and Pentodes 53
  4.3 Relation between Signal to Noise Ratio, Resolving
      Time and Amplifier Bandwidth 59
  4.4 Choice of the Input Valve and Operating Conditions 67

		PAGE
4.5	Best Possible Signal to Noise Ratio	68
4.6	Minimising Effects due to Variation in Collection Time	69
4.7	Pulse Shaping Using a Critically Damped Ringing Coil	71
4.8	Pulse Shaping Using a Shorted Delay Line	72

5 SENSITIVITY 79
| 5.1 | Amplifier Sensitivity in Energy Measurements | 79 |
| 5.2 | Amplifier Sensitivity in Total Activity Measurements | 86 |

6 PROPORTIONAL AND SCINTILLATION COUNTERS 105
6.1	Proportional Counters	105
6.2	Experimental Determination of the Collection Time of a Proportional Counter	108
6.3	Proportional Counting in the Presence of an Intense Background of Weakly Ionising Particles	111
6.4	Scintillation Counters	120
6.5	Scintillation Counter Bandwidth for Optimum Signal to Noise Ratio	123

REFERENCES	127
APPENDICES	129
INDEX	153

EDITOR'S PREFACE

The aim of these monographs is to report upon research carried out in electronics and applied physics. Work in these fields continues to expand rapidly, and it is recognised that the collation and dissemination of information in a usable form is of the greatest importance to all those actively engaged in them. The monographs will be written by specialists in their own subjects, and the time required for publication will be kept to a minimum in order that these accounts of new work may be made quickly and widely available.

Wherever it is practicable the monographs will be kept short in length to enable all those interested in electronics to find the essentials necessary for their work in a condensed and concentrated form.

D. W. FRY

AUTHOR'S PREFACE

The tremendous war time advances in nuclear science and the subsequent governmental and industrial expansion of research and development work in this field have created a greatly increased demand for electronic amplifiers as used with nuclear counters.

Such amplifiers do not differ appreciably in basic circuit design from those used in television and radar, and this aspect of the subject is covered to a large extent by the published literature. The factors influencing the signal to noise ratio and resolution in nuclear counter amplifiers are however not so well understood or publicised (see references 1 to 6 inclusive at the end of the text) and it was a realisation of this need which prompted the author to write the present monograph.

In preparing the text, the author has endeavoured to give a

comprehensive, yet essentially simple treatment of a far from simple subject. This has necessarily involved the inclusion in Chapters 1, 2 and 6 of material relating to the operation of nuclear counters which has already been published elsewhere. Chapters 3, 4 and 5, however, on noise, signal to noise ratio, resolution and sensitivity contain much theoretical and experimental data which has not been published before, and in addition the analysis given in Chapter 5 of the sensitivity of amplifiers used for total activity measurements, together with its experimental support, is believed to contain many original features.

The book will appeal in the main to two classes of reader. Firstly to the young graduate, about to enter the field of nuclear science for the first time and in search of a compact and up to date presentation of the subject of signal, noise and resolution in nuclear counter amplifiers. Secondly to the experienced nuclear physicist or radiochemist, interested in the sensitivity and speed limitations of such amplifiers and in how to achieve the optimum performance from his equipment in a particular experiment. At the same time it is hoped that much of the general information on noise and signal to noise ratio will prove of interest to a considerably larger body of readers.

The author would like to express his indebtedness to Dr J. HOWLETT, Mr W. WALKINSHAW, and Mr P. M. WOODWARD. To the former for computing many of the points in the theoretical curves and to all three for stimulating discussions on general problems of signal to noise ratio. Thanks are also due to Mr K. KANDIAH for having read the manuscript and having made many valuable suggestions, and to Dr D. TAYLOR for his constant support during the writing of the text.

A. B. G.

ABINGDON
January 1953

LIST OF SYMBOLS
(in order of usage in the text.)

C	chamber capacity, farads
k	BOLTZMANN's constant, joules/°K
T_e	absolute temperature, °K
e	charge on electron, coulombs
D	distance between plates of ionisation chamber, cm
d	distance between ionisation and collector plate, cm
T	collection time, sec
N_p	total number of primary ion pairs
R	chamber feed resistor, ohms
$\bar{n}$	mean occurrence rate of signal pulses, pulses/sec
δt	small interval of time, sec
$p_{1.2.3.\ \text{etc.}}$	probability of 1 2 3 etc. pulses occurring in time δt
V	voltage across chamber, volts
V_g	grid voltage, volts
g	region of ionisation in gridded chamber
h	region between grid and collector plate
ΣC	total input capacity, farads
$T_1 = C_1 R_1$	amplifier differentiating time constant, sec
$T_2 = C_2 R_2$	amplifier integrating time constant, sec
G	amplifier gain
Q	signal charge, coulombs
V_0	pulse voltage at chamber, volts
ε	base of Napierian logarithms
t	time variable, sec
$V_{(0-T)}$	pulse voltage during time 0 to T, volts
$V_{(T-\infty)}$	pulse voltage during time T to ∞, volts
∞	infinity
V_m	maximum pulse voltage, volts
T_r	resolving time, sec

LIST OF SYMBOLS

$\dfrac{v^2}{\delta f}$	spectral density of noise voltage, volts^2-sec
$\dfrac{i^2}{\delta f}$	spectral density of noise current, amp^2-sec
v_t	r.m.s. thermal noise voltage, volts
f	frequency, cycles/sec
π	constant
$\omega = 2\pi f$	angular frequency, radians/sec
f_1	lower frequency limit, cycles/sec
f_2	upper frequency limit, cycles/sec
i_s	r.m.s. shot noise current, amp
I_a	anode current, amp
F^2	space charge smoothing factor
g_m	mutual conductance, amp/volt
v_s	r.m.s. shot noise voltage, volts
I_s	screen current, amp
I_c	cathode current, amp
i_g	r.m.s. grid current noise current, amp
I_g	grid current, amp
I_{g+}	electron grid current, amp
I_{g-}	reverse grid current, amp
v_g	r.m.s. grid current noise voltage, volts
V_a	anode voltage, volts
V_s	screen voltage, volts
V_h	heater voltage, volts
v_f	r.m.s. flicker noise voltage, volts
τ	electron transit time, sec
J_1	constant
R_a	anode resistor, ohms
μ	amplification factor
ρ	anode impedance, ohms
J_2	constant
I	current through resistor, amp
A_1	particular pulse amplitude at amplifier output, volts

LIST OF SYMBOLS

E_1	energy corresponding to A_1, eV
x	displacement from centre of Gaussian distribution, volts
v	r.m.s. noise voltage, volts
S_1	signal level ⎫
S_2	signal level ⎬ relating to counting plateau, volts
N_o	noise level ⎭
C_g	grid to cathode capacity, farads
C_s	screen to grid capacity, farads
C_a	anode to grid capacity, farads
v_{pen}	r.m.s. shot noise voltage of pentode, volts
$\dfrac{S}{N}_{(\text{shot})}$	signal to shot noise ratio
v_{tri}	r.m.s. shot noise voltage of triode, volts
M	stage gain
v_a	r.m.s. shot noise voltage on anode, volts
n	number of valves in parallel
$\dfrac{S}{N}_{(\text{grid current})}$	signal to grid current noise ratio
$\dfrac{S}{N}_{(\text{flicker})}$	signal to flicker noise ratio
θ	ratio T_1/T_2
X	constant relating $[T_1 T_2]^{1/2}$ to T
t_m	time at which pulse maximum occurs, sec
ϕ	constant relating two $\dfrac{I_g}{g_m}$ factors
L_d	inductance of ringing coil, henrys
C_d	capacity of ringing coil, farads
R_d	resistor for critical damping of ringing coil, ohms
T_d	delay time of line, sec
Z_0	characteristic impedance of line, ohms
C_t	test capacity, farads
V_t	voltage of test waveform, volts
$\overline{N}$	mean occurrence rate of noise events, pulses/sec
$V(t)$	voltage effect produced by single noise event, volts

LIST OF SYMBOLS

$p_{(x, x-dx)}$	probability of noise waveform being within limits x and $x - dx$
$\dot{V}(t)$	differential of $V(t)$
$p_{(\text{vel}+)}$	probability of noise waveform having positive velocity
p_x	probability of noise waveform passing through x with positive velocity
P	rate of counting of noise crests, counts/sec
P_g	rate of counting of grid current noise crests, counts/sec
P_s	rate of counting of shot noise crests, counts/sec
Δ	amplifier response time, sec
Δ_g	effective response time for grid current noise events, sec
Δ_s	effective response time for shot noise events, sec
z	ratio of discriminator level to 4.65 times the r.m.s. grid current noise
r	arbitrary radius in proportional counter, cm
a	radius of centre wire, cm
b	radius of outer electrode, cm
m	number of mean free path lengths
M_g	gas multiplication factor
H	amplitude ratio between wanted pulses and background pulses
p_H	probability per pulse of background fluctuation H times amplitude of single pulse
P_H	rate of counting of background fluctuations exceeding H times amplitude of single pulse, counts/sec

1
INTRODUCTION

A NUCLEAR counter can be defined as any device which will detect and count individual charged particles arising from radioactive disintegrations. The nuclear counters in most frequent use today are the ionisation chamber, the proportional counter, the GEIGER-MÜLLER counter and the scintillation counter. The G.M. counter, in so far as the present monograph is concerned, is in a class by itself. It gives an output pulse whose amplitude is not dependent on the energy expended by the charged particle in the counter, it requires practically no valve amplification, and it can not be used at very high counting rates due to the long dead time which follows each pulse. For these reasons the G.M. counter is not discussed in the text.

Of the other counters mentioned above, the ionisation chamber and the proportional counter are both gas ionisation devices in which the energy of a high speed charged particle is expended in producing ionisation of the gas inside the counter. On the average an energy of 30 electron-volts (eV) is required for each ion pair produced. The electronic charge liberated in this ionisation process is swept to the counter plates by an artificially established electrostatic field and produces a small voltage step or pulse across the counter capacity. The proportional counter is a variation of the ionisation chamber in which secondary ionisation is made to occur by the use of a much higher field strength. This results in a substantially increased output pulse having an amplitude proportional to the number of primary ion pairs produced.

The scintillation counter on the other hand detects a high speed charged particle by the light flash which it produces in a phosphor. The energy of the charged particle is expended in producing photons in the phosphor, the energy required per photon being of the same order of magnitude as the energy expended per ion pair in a gas. The photons are converted to photo-electrons by the light sensitive cathode of a multiplier photocell and this current pulse in turn produces a voltage pulse across the load in the photomultiplier final anode.

The high speed charged particles which are detected by the above nuclear counters may be the primary products of a radioactive disintegration, the classical examples being α particles and β particles. Alternatively they may be secondary particles which receive energy from the primary constituents, examples here being electrons from X- and γ-rays and protons from fast neutrons. Then again some of the primary constituents may be used to initiate a second nuclear reaction in which charged particles are produced. These charged particles are then detected by the nuclear counter. Examples are α particles from slow neutrons and fission fragments from slow or fast neutrons.

For accurate quantitative measurements, the ionisation chamber has led the field for a large number of years. More recently however the proportional counter has become popular and is now used for many counting applications which were previously carried out by ionisation chambers. The popularity of the proportional counter is due primarily to the increased output pulse which gives a considerable measure of immunity from amplifier noise, although the secondary ionisation process does introduce additional problems of a different nature. In still more recent years much attention has been given to exploiting the versatility of the scintillation counter. The practical application of this type of counter has been made possible almost entirely by the development of the modern high gain photomultiplier.

It is true however that for many applications, particularly where energy discrimination is required, the ionisation chamber reigns supreme. For this reason, together with the fact that the amplifier noise is of most consequence when using a counter of this type, the main body of the text is devoted to considerations of signal to noise problems in amplifiers for use with ionisation chambers. This emphasis is later justified, however, because it is shown in Chapter 6 that a firm understanding of ionisation chamber techniques is sufficient to deal with all problems of a similar nature which arise from the use of proportional and scintillation counters.

The function of the pulse amplifier is to accept the small voltage impulses produced by the nuclear counter and to amplify these pulses to a level usually in the region of 5 to 50 volts, at which level the pulses are suitable for driving scalers and registers, ratemeters or kicksorters. These latter instruments carry out measurements of pulse rate and pulse amplitude distribution and from them, information can be obtained relating to the amount of radioactive material present and to the energies of the emitted radiations. From a practical point of view the only design features of the pulse amplifier which enter into subsequent discussions are the bandwidth and the noise characteristics of the valve and components used for the input stage. No details are given of the type of thermionic valve circuits which follow the input stage—this is a subject quite outside the scope of the present monograph. The reader should appreciate however that the amplifiers under consideration are all regarded as being designed around conventional receiving valves, covering a total frequency range of from a few hundred cycles per second to several megacycles per second, and having voltage gains ranging from a maximum of about 10^3 for scintillation counters to about 10^6 for ionisation chambers.

2
NATURE OF THE SIGNALS FROM IONISATION CHAMBERS

2.1 ION CHAMBERS

The simplest of all nuclear counters is the parallel plate air*-filled ion chamber. This counter consists basically of two parallel metal plates, separated by air and having an inherent self capacity C hereafter known as the chamber capacity. When a high speed charged particle, arising from a radioactive disintegration, passes through the space between the plates the air molecules along its track are ionised, the particle losing approximately 30 eV of energy for each ionising collision. Positive ions and electrons are produced but because of the air filling of the chamber, the electrons are immediately captured by oxygen atoms forming negative ions. In the absence of an electrostatic field between the plates, the ions move at random with an average energy of agitation equal to $\frac{3kT_e}{2}$ joules and will eventually recombine after a time of some 300 sec[7]. In the above expression

$k =$ BOLTZMANN's constant $= 1 \cdot 37 \times 10^{-23}$ joules/°K

$T_e =$ absolute temperature.

The application of a potential difference between the plates from a battery or similar voltage source and the consequent establishment of an electrostatic field gives the ions a drift component of velocity towards the plates, the positive and negative ions moving of course in opposite directions. As the applied potential is increased the ionic recombination diminishes until above a critical value known as the saturation voltage of the chamber, no recombination occurs and all the ions formed are collected on the plates. The drift velocity is determined by an equilibrium condition between the

---

* Ion chambers having gas fillings other than air are not described in this section as such chambers are rarely used and offer no advantages over the simple type which is open to the atmosphere.

energy supplied by the electrostatic field to the ions per unit time and the energy lost through collisions with neutral gas molecules. The drift velocity is thus proportional to the intensity of the electrostatic field and inversely proportional to the gas pressure. In air at N.T.P. the drift velocity of positive and negative ions is approximately 1·3 cm/s for a field intensity of 1 volt/cm.

Suppose one plate of the chamber is effectively earthed and consider the charge conditions of the other plate, hereafter referred to as the collector plate. At the instant of formation of the ions and provided the field intensity is uniform, as is the case in a parallel plate chamber, a single ion pair induces equal positive and negative charges of magnitude $\dfrac{e(D-d)}{D}$ on the collector plate, where

e = charge on an electron = $1·59 \times 10^{-19}$ coulombs

d = distance of ion pair from collector plate

D = distance between plates.

As the ions move apart the induced charge due to the one moving towards the collector plate increases linearly to the ultimate value e at collection while that due to the receding ion decreases linearly to zero. Eventually, after a time T known as the collection time of the chamber, all the ions reach the plates and if N_p is the total number of ion pairs initially produced, the final charge on the collector plate is $N_p e$ coulombs.

If the battery is connected to the collector plate through a large resistor R so that the time constant CR is many times greater than the collection time T, the charge $N_p e$ produces a small voltage step on the collector plate of magnitude $N_p e/C$. This voltage step then decays to zero with a time constant CR as the charge leaks away through the resistor R.

An ion chamber working under such conditions is shown in fig. 1(a), and fig. 1(b) shows a typical voltage pulse developed on the collector plate.

It will be appreciated

(a) That the appearance of a voltage pulse on the collector plate is indicative of the presence of an ionising particle inside the chamber.

(b) If the ionising particle loses all its energy producing ionisation inside the chamber, the amplitude of the voltage pulse is a measure of the initial energy of the particle.

It is the function of the pulse amplifier to accept the small ion chamber pulses, which may be of the order of a few microvolts to a few hundred microvolts, and to amplify them to a level usually between 5 and 50 volts, at which level, measurements of pulse rate and pulse amplitude can be carried out by other instruments to a high degree of accuracy.

The pulses fed to the amplifier from the ion chamber occur at random in time. If the mean rate of occurrence is $\bar{n}$ per second,

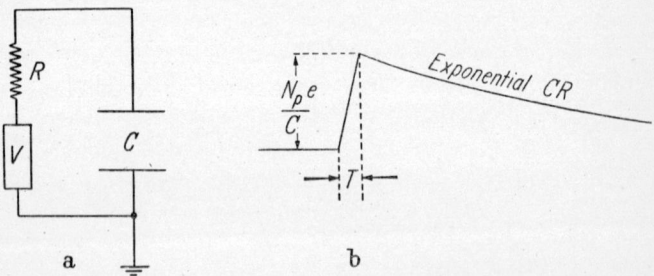

Fig. 1. Circuit arrangement of ion chamber with typical voltage pulse produced on collector plate

then the probability of a pulse occurring in a small interval of time δt seconds is given by POISSON's formula and is

$$p_1 = \varepsilon^{-\bar{n}\,\delta t}\frac{(\bar{n}\,\delta t)^1}{1!}.$$

If δt is small compared with the mean period between pulses, i.e. $\bar{n}\,\delta t \ll 1$, this reduces to

$$p_1 \doteqdot \bar{n}\,\delta t.$$

Similarly the probability of two pulses occurring in the time δt is

$$p_2 = \varepsilon^{-\bar{n}\,\delta t}\frac{(\bar{n}\,\delta t)^2}{2!}$$

$$\doteqdot \frac{(\bar{n}\,\delta t)^2}{2} \text{ for } \bar{n}\,\delta t \ll 1.$$

If the interval δt is considered as commencing immediately after the start of any pulse, then $\bar{n}\,\delta t$ can also be regarded as the probability per pulse of a second pulse occurring within this interval. This will happen $(\bar{n})^2 \delta t$ times per second, and thus the percentage of pulses which follow previous pulses within a time interval δt is simply given by $100\,\bar{n}\,\delta t\%$.

From the point of view of accepting pulses spaced randomly in time, and particularly at a high mean recurrence rate, the long exponential tail of each pulse becomes an embarrassment in so far as a considerable proportion of subsequent pulses will occur before previous pulses have decayed to zero. This will clearly give rise to errors in pulse amplitude measurement, and errors in pulse rate can also occur due to closely overlapping pulses being recorded by the measuring instruments as a single pulse. To keep such errors small the duration of the pulse at the amplifier output should be as short as possible. The ability of a chamber and amplifier to accept pulses at a high mean rate and at the same time keep amplitude and pulse rate errors at a low value is a measure of the resolving power of the system.

To improve the resolving power of a chamber and amplifier combination it is permitted to incorporate somewhere in the system a pulse shaping network whose prime function is to cause the trailing edge of the pulse to fall as rapidly as possible after the collection time T of the chamber. It is also permissible for the pulse shaping network to cause considerable distortion of the pulse front, provided the relationship between pulse amplitude and particle energy which exists at the collector plate is preserved at the amplifier output.

The simplest and most commonly used pulse shaping network is the short time constant or so called differentiating circuit which comprises only two components—a condenser and a resistor. Other forms of pulse shaping network, notably the critically damped ringing coil and the shorted delay line, are less versatile and of more limited application. The analyses contained in this and later chapters, are devoted almost entirely to pulse shaping using the condenser resistor differentiating circuit. The critically damped ringing coil and the shorted delay line are discussed only in Chapter 4 where the former is shown to be comparable with, and the latter slightly superior to, the differentiating circuit as regards resolving power for a given signal to noise ratio.

The application of a differentiating circuit to the shaping of an ion chamber pulse is shown in fig. 2.

In the ion chamber and amplifier system at present under discussion, the pulse may be shaped either

    (a) at the chamber prior to its passage through the amplifier
or (b) at some stage within the amplifier.

In method (a) the chamber capacity C and the feed resistor R actually form the differentiating circuit provided R is very much reduced in magnitude. Reduction of R causes the trailing edge of the pulse to fall more rapidly, with consequent narrowing of the pulse width and an overall improvement in the resolving power of

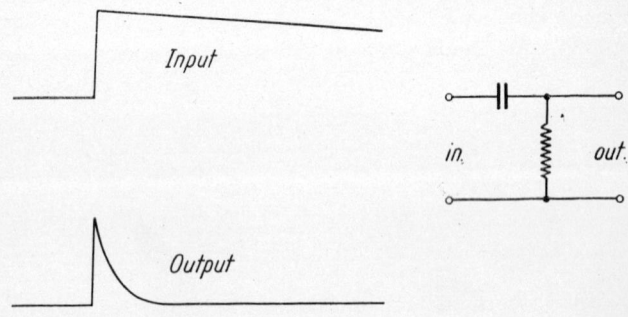

Fig. 2. Use of a differentiating circuit to shorten duration of a long pulse

the system. In the limit when R has fallen to such a value that the time constant CR is comparable with the chamber collection time T, further reduction of R then causes the pulse amplitude to drop rapidly with little or no reduction in pulse width. In this simple analysis, therefore, it will suffice to say that the differentiating time constant may be made as short as the chamber collection time, but making it very much shorter produces no substantial improvement in the resolving power of the system. In fact, at this stage, the collection time of the counter is the limiting factor.

In method (b) the feed resistor to the chamber is kept large and a separate differentiating circuit is incorporated at some stage in the amplifier. Provided no overloading occurs in the amplifier prior to the differentiating circuit, the effect on pulse amplitude and pulse width is exactly the same as that discussed for method (a) above.

In amplifiers for use with ion chambers it is the invariable rule to employ method (b). This is so, because, as will be shown in

Section 3.2, method (b) is effective in reducing the noise contributions from R and from the first valve in the amplifier whereas method (a) is not. This is of prime importance in ion chamber amplifiers where the chamber voltage pulses are usually very small. It will be appreciated that when method (b) is employed there will be considerable superposition of pulses in the stages of the amplifier prior to the differentiating time constant. Fig. 3 shows typical waveforms before and after differentiation, due to the succession of random pulses from a radioactive source.

The voltage fluctuations before differentiation are appreciably greater than those after differentiation, and as a consequence the

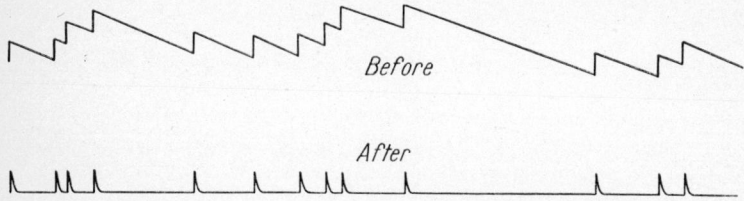

Fig. 3. Signal waveform before and after differentiation

position of the differentiating time constant in the amplifier must be carefully chosen in order to prevent overloading of the early stages, especially when the mean recurrence rate of the pulses is very high.

It now only remains to discuss the shape of the leading edge of the chamber pulse. This rise is not instantaneous, when the ions reach the plates, but is gradual and occurs during the time that the ions are in motion. So far, in this simplified analysis, the leading edge of the chamber pulse has been illustrated diagramatically as having a constant rate of rise. This is a case which can occur in practice and corresponds to the ionisation taking place very close to one of the chamber plates. The ions of one polarity are collected practically instantaneously on the near plate without contributing anything to the pulse amplitude, while those of the other polarity drift towards the distant plate causing the pulse on the collector to rise linearly to the value $N_p e/C$. The collection time T for this particular case corresponds to the time taken by an ion to drift the distance between the plates.

When the ionisation is not localised near one of the plates, but occupies some intermediate position, the leading edge of the pulse departs from the simple linear case described above although the final amplitude remains the same. The collection time may also be different—it can be shorter than the value for T defined above but it can never be longer.

To analyse theoretically the effect of the differentiating time constant on the chamber pulse it is necessary to assume some simple shape for the latter. Most writers prefer to consider the chamber pulse as a step function as this considerably eases the labour involved in an analysis of this kind. The author feels, however, that some allowance should be made for the finite rise time, especially as in many present day uses of pulse amplifiers the differentiating time constant and the chamber collection time are of the same order. Accordingly, the simple linear rise has here been chosen as standard and is used in most analyses in this and later chapters.

The air-filled ion chamber was used a great deal in the early days of nuclear research, but its present day use is very limited. It has been described in detail, however, because it is the simplest nuclear counter to understand, it gives the reader an introduction to the concept of resolving power of a chamber and amplifier system, and its mechanism of pulse formation can readily be applied to most modern counters. The main disadvantage of the air-filled chamber is its long collection time, which limits resolution and prohibits counting at high rates or the discrimination against very high unwanted backgrounds. For example, with a field intensity of 1000 volts/cm the drift velocity of the ions is approximately 1300 cm/s and for average size chambers this gives collection times of the order of one half to several milliseconds. Modern counters, to be described later in the text, have collection times of 1000 to 10,000 times shorter.

2.2. ELECTRON CHAMBERS

Basically the simple electron chamber is similar to the ion chamber just described, with the exception that the air filling of the latter is replaced by a filling of a gas having no affinity for electron capture. Typical examples are hydrogen, nitrogen, helium, neon, argon and carbon dioxide. Accordingly, in a chamber of this type, the electrons liberated by the ionising particle remain as free electrons and are eventually collected on the positive plate. On

NATURE OF THE SIGNALS FROM IONISATION CHAMBERS

the average, the drift velocity of electrons in a particular parent gas is about one thousand times higher than the drift velocity of the corresponding positive ions. The electrons are therefore collected in a very short time, of the order of 1 microsec. for an average size chamber, while the positive ion collection takes a time of the order of 1 millisec. It can in fact be assumed that the electron collection is virtually complete before the positive ions begin to move.

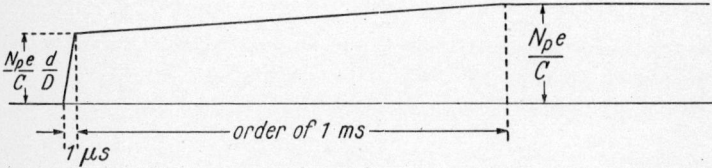

Fig. 4. Voltage pulse produced by simple parallel plate electron chamber

If the collector plate of the chamber is positive, it attracts the electrons to it while the positive ions slowly drift towards the earthy plate. If the ionisation occurs at an average distance d from the collector plate, the charge on the latter, after complete collection of the electrons, is

$$N_p e - N_p e \frac{(D-d)}{D}$$
$$= N_p e \frac{d}{D}.$$

The first term represents the charge due to the electron collection and the second term represents the induced charge of the positive ions. This phase occupies a time of the order of 1 microsec. As the positive ions drift slowly towards the earthy plate the charge slowly increases from $N_p e/(d/D)$ to the ultimate value $N_p e$, the time interval here being approximately 1 millisec. The voltage pulse produced on the collector plate is shown in fig. 4*.

---

* In the above figure the representation is that of a positive pulse. The reader will appreciate however that in this, and in fact in all counters where electrons or negative ions reach the collector plate, a negative pulse will be produced. For the sake of uniformity in the diagrams and to avoid confusion in the text it is proposed hereafter to illustrate all counter pulses in the positive sense irrespective of their actual phase.

If the pulse on the collector plate is differentiated by a time constant comparable with the collection time of the electrons the resulting pulse appears as in fig. 5.

The response is due almost entirely to the fast rise resulting from electron collection with little or no contribution from the rise due to the positive ions. Hence, by using a chamber of this type followed by a very short differentiating time constant, the resolving power

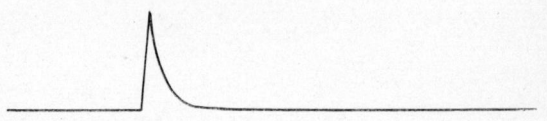

Fig. 5. Differentiated pulse from simple electron chamber

of the counting system can be enormously improved. The disadvantage of this simple counter, however, is that the differentiated pulse amplitude is not only proportional to the energy expended by the ionising particle inside the chamber, but is also a function of the distance from the collector plate at which the ionisation occurs.

This simple chamber has therefore little practical value, but this serious limitation can be easily removed by the introduction of a wire mesh or FRISCH grid[8] between the source of the ionisation and the collector plate. Such a gridded electron chamber is shown diagramatically in fig. 6.

In operation it is arranged that all the ionisation occurs in the region g between the earthy plate and the grid. The grid is held at an intermediate positive potential V_g with respect to the earthy plate. The collector plate is, as before, polarized through the large resistor R. When ionisation is produced in the region g, the electrons drift towards the grid and the positive ions in the opposite direction. During this phase, no charge is produced on the collector plate since this is electrostatically screened by the grid from the induced effect of the electrons and the positive ions. When the electrons reach the grid, however, because of its open mesh construction, they pass through into the region h, with only a very small percentage of the electrons suffering capture by the grid. In this region the electrons begin to induce on the collector plate a charge which rises from zero to the full value $N_p e$ as the electrons move through the distance from the grid to the collector plate. In the same time the voltage pulse rises to the amplitude $N_p e/C$ and is thereafter uninfluenced by the positive ions left in the region g.

The gridded chamber is thus a very practical form of the simple electron chamber, enabling the high resolving power associated with electron collection to be maintained and at the same time preserving the relationship between pulse amplitude and particle energy.

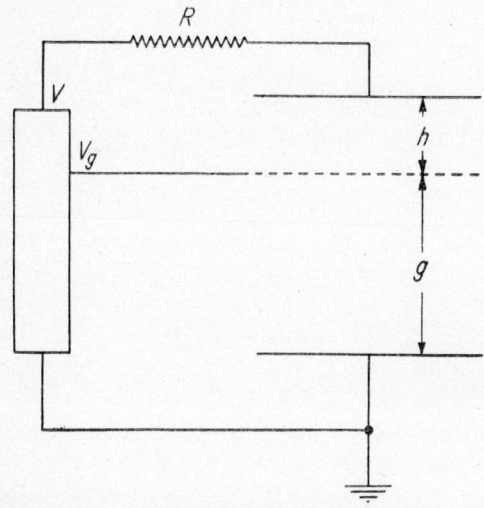

Fig. 6. Circuit arrangement of gridded electron chamber

If the ionisation is localised at any point between the grid and the earthy plate, or is distributed along any line parallel to the grid, the electrons all pass through the grid at the same instant and the pulse on the collector plate then rises linearly to the value $N_p e/C$ in the time taken by a single electron to drift the distance between the grid and the collector plate. Like the simple ion chamber, this is a condition which can easily be made to occur in practice. If the ionisation is distributed along a line having a component perpendicular to the grid, the leading edge of the pulse departs from the simple linear rise and the collection time tends to increase. In the limit, however, the collection time can never exceed the time taken by a single electron to drift the total distance between the earthy and the collector plates and this is a case which very rarely occurs in practice.

From the point of view of assuming some simple shape for the leading edge of the chamber pulse the most obvious choice is as before, the linear rise to the value $N_p e/C$. It will be appreciated therefore that the same calculations are applicable to both ion chambers and gridded electron chambers provided the appropriate collection time is used in each case.

2.3. EFFECT OF AMPLIFIER TIME CONSTANTS ON PULSE AMPLITUDE AND PULSE WIDTH

In this section it is proposed to analyse theoretically the effect which the finite pass band of the amplifier has on the shape of the ingoing pulse. The analysis is applicable to both ion and electron

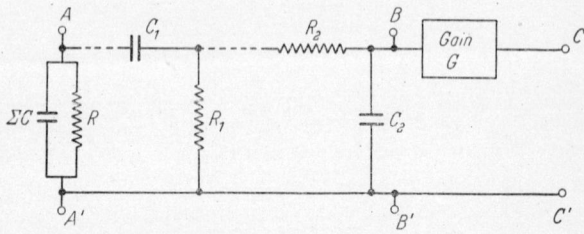

Fig. 7. Equivalent circuit of ionisation chamber and pulse amplifier

chambers since the pulse shapes are similar except for a difference in time scale. It is convenient to assume an equivalent circuit for the amplifier and chamber combination and this is shown in fig. 7.

The input terminals of the amplifier are AA', ΣC represents the total input capacity due to the chamber, amplifier and any connecting lead and R represents the feed resistor to the chamber which is effectively in shunt with ΣC as far as pulse voltages are concerned. The time constant of differentiation is $C_1 R_1 = T_1$ and it is assumed that the top frequency limit of the amplifier is determined by a single time constant of integration $C_2 R_2 = T_2$. The dotted connections between ΣCR and T_1 and between T_1 and T_2 imply that these circuits are quite independent and do not impose electrical loading on each other. The gain G represents the valve magnification of the amplifier and is of course uniform over the whole frequency spectrum, since the finite limits of the pass band are determined in the equivalent circuit by T_1 and T_2.

NATURE OF THE SIGNALS FROM IONISATION CHAMBERS 15

The analysis is based on the assumption that the input pulse at the terminals AA' rises linearly to a voltage $V_0 = Q/\Sigma C$ where $Q = N_p e$ is the total charge collected in the chamber, reaching this amplitude in a time T equal to the collection time of the chamber and then remaining substantially at this amplitude for a time which is long compared with the differentiating time constant T_1. This pulse is shown in fig. 8.

It is required to determine an expression or expressions for the

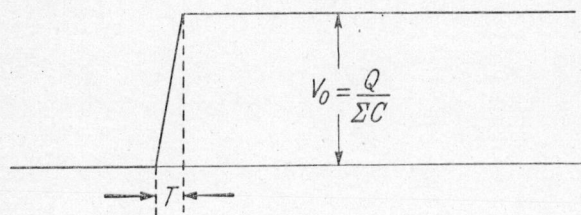

Fig. 8. Theoretical voltage pulse from ionisation chamber

pulse reaching BB' in the equivalent circuit. The pulse at the output terminals CC' will be similar, but G times greater. The analysis is contained in Appendix A and the following equations, of interest in the present discussion, are derived.

A general expression for the pulse at BB' and valid over the time interval 0 to T is given by

$$V_{(0-T)} = \frac{V_0 T_1}{T}[1 - \varepsilon^{-t/T_2}] - \frac{V_0 T_1^2}{T(T_1 - T_2)}[\varepsilon^{-t/T_1} - \varepsilon^{-t/T_2}]. \quad (1)$$

A similar expression but valid over the time interval T to ∞ is given by

$$V_{(T-\infty)} = \frac{V_0 T_1^2}{T(T_1 - T_2)}[\varepsilon^{T/T_1} - 1]\,\varepsilon^{-t/T_1}$$

$$- \frac{V_0 T_1 T_2}{T(T_1 - T_2)}[\varepsilon^{T/T_2} - 1]\,\varepsilon^{-t/T_2} \quad (2)$$

In subsequent discussions on signal to noise ratio it is necessary to know the maximum value of the pulse reaching BB' in the equivalent circuit in order that this can be compared with noise levels at the same point. This is also derived in Appendix A and is given by

$$V_m = \frac{V_0 T_1}{T} \frac{\left[\varepsilon^{T/T_1} - 1\right]^{\frac{T_1}{T_1-T_2}}}{\left[\varepsilon^{T/T_2} - 1\right]^{\frac{T_2}{T_1-T_2}}}. \qquad (3)$$

Equation (3) is a very important one and is represented graphically by the series of curves shown in fig. 9.

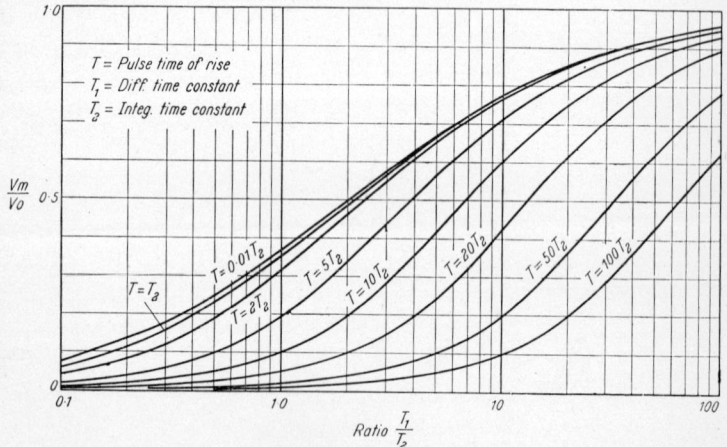

Fig. 9. Pulse amplitude as a function of chamber collection time and amplifier time constants

Each curve refers to a particular value of the ratio T/T_2 and has been plotted for a wide range of the variable T_1/T_2 (T_2 constant, of course, for any given curve). Each curve shows therefore, for a fixed relation between the chamber collection time and the amplifier integrating time constant, how the pulse amplitude at BB' alters with variation of the differentiating time constant. The curves have many applications and are frequently referred to in later chapters.

The other important criterion which comes into subsequent discussions is the resolving time T_r of the chamber and amplifier combination. The pulse at the amplifier output is normally fed to an amplitude discriminator which is a circuit that responds to all pulses above a certain voltage level and disregards all pulses below this level. When a pulse passes the threshold level of the discriminator, the latter, which is usually a trigger circuit, fires and produces an output pulse of fixed amplitude and fixed width and which is suitable for driving a scaler and mechanical register or a ratemeter. The amplitude discriminator is thus the basis of the instrument which measures pulse amplitude at the amplifier output while the scaler and register or the ratemeter deal with counting rate measurements.

The resolving time of the chamber and amplifier combination is the time from the start of a pulse which triggers the discriminator,

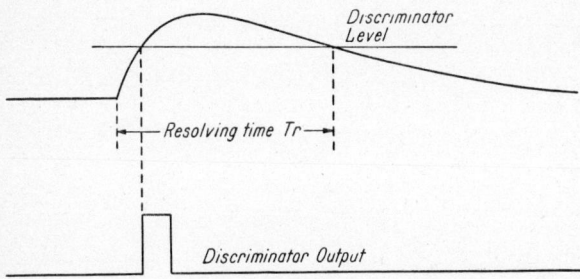

Fig. 10. Resolving time of pulse amplifier used with amplitude discriminator

until the trailing edge falls again below the threshold level, as no subsequent pulse which starts within this time interval is recorded by the measuring instruments as a separate pulse. This is illustrated in fig. 10.

The resolving power of the chamber and amplifier combination, mentioned frequently in Sections 2.1 and 2.2 is clearly the reciprocal of the resolving time T_r as now defined. The resolving time is dependent on the setting of the discriminator level as well as on the collection time T and the amplifier differentiating and integrating time constants T_1 and T_2 respectively. It is not possible to derive a reasonable expression which specifies this duration accurately as a function of all four variables, so for the time being it is proposed to introduce a simplified definition of resolving time, which will suffice until some further results have been obtained. The resolving time will hereafter be regarded as the average width of the amplifier pulse, obtained by dividing the area of the pulse by the maximum amplitude. Equation (3) gives the maximum amplitude and the area is obtained by integrating equation (1) from 0 to T and equation (2) from T to ∞ and summing the results. The integrations are worked out in Appendix B, and the area is found to be simply $V_0 T_1$ for all values of T T_1 and T_2.

The resolving time is therefore given by

$$T_r = \frac{V_0 T_1}{V_m}$$

$$= \frac{T \left[\varepsilon^{T/T_2} - 1 \right]^{\frac{T_2}{T_1 - T_2}}}{\left[\varepsilon^{T/T_1} - 1 \right]^{\frac{T_1}{T_1 - T_2}}}. \qquad (4)$$

3
VALVE AND CIRCUIT NOISE

3.1. GENERAL

In an electrical conductor which is not in connection with a source of external voltage, the free electrons in the conductor are in a constant state of thermal agitation. The motion of the electrons is quite random and corresponds to small fluctuations of charge and as a consequence small fluctuations of current within the conductor. Such current fluctuations in turn give rise to voltage fluctuations across the ends of the conductor, the magnitude of the voltage fluctuations depending on the resistance of the conductor. If measured over an infinitely long period of time, the voltage and current fluctuations must be zero otherwise there would be a unidirectional flow of current in the conductor, which is clearly impossible since the ends are free. This phenomenon is known as thermal noise[9,10].

When a steady current flows in the conductor under the influence of an externally applied voltage, the free electrons then acquire a drift component of velocity in addition to their thermal agitation velocities. For all normal currents, however, this drift velocity is very much smaller than the average thermal agitation velocity of the electrons and thus the thermal noise generated is not dependent on the magnitude of any steady current flowing in the conductor.

Similarly, in a thermionic valve where the passage of current is due to the transfer of discreet units of charge from a cathode to a collecting electrode, the mean current passed has superimposed fluctuations due to the random emission of electrons from the cathode and the magnitude of these fluctuations depends on the magnitude of the mean current. This phenomenon is known as the shot effect[9,10] in valves and is another source of noise.

Most forms of noise in thermionic valves and electrical circuits can usually be attributed to one or other of the above basic phenomena. Although the noise voltage or current fluctuations must be zero if measured over an infinitely long period of time, the amplifiers discussed in the text are effectively making measurements over short time periods, and noise fluctuations must therefore be considered. The time during which a practical amplifier effects a measurement of some discontinuous input event, such as a pulse, is clearly related to the resolving time of the amplifier and this in turn is dependent on the bandwidth. In consequence, the noise fluctuations will likewise depend on the amplifier bandwidth.

The sinusoidal components which can be considered as making up the noise fluctuations are distributed throughout the frequency spectrum and are moreover quite random in phase. Nevertheless, there is a definite relationship between the magnitude of the noise fluctuations and the power dissipated in the system, and it is thereby customary to measure the noise fluctuations in terms of power or mean square voltage or mean square current. Since the amplifiers under consideration are fundamentally voltage amplifiers it is the object of subsequent analyses to express the noise fluctuations in terms of the equivalent mean square or root mean square (r.m.s.) voltage. It should be noted that the combination of two or more noise voltages is achieved by adding arithmetically their mean square values.

In addition to classifying noise into the two main divisions described above, it is usually possible to further subdivide by specifying how the noise components are distributed in the frequency spectrum. This classification describes the spectral density of the noise and is expressed by the mean square voltage or the mean square current per unit bandwidth—that is $v^2/\delta f$ or $i^2/\delta f$ respectively.

The minimum signal which can be usefully amplified by any amplifier is limited by the magnitude of the noise, since a signal which is small compared with the noise will not be discernible or measureable at the amplifier output. All valves and resistors in an amplifier generate noise but in general the limiting sensitivity is determined by the noise voltages arising from the first valve and associated components, as such voltages are subject to the full magnification of the amplifier.

For the calculation of noise voltages, the equivalent circuit of fig. 7 is used with the added assumption that the grid of the first valve is considered as being connected to the input terminal A. This simply means that all noise voltages due to the valve and which can be expressed as equivalent voltages at the grid, are treated as being effective at point A. The object of the analyses in this chapter is

3.2. THERMAL NOISE

As described in the previous section, this type of noise arises from the thermal agitation of the free electrons in an electrical conductor.

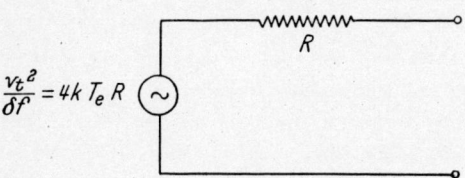

Fig. 11. Representation of thermal noise by equivalent noise generator

Its voltage spectral density is constant and is given by the following equation[11, 12].

$$\frac{v_t^2}{\delta f} = 4kT_eR, \qquad (5)$$

where

$k =$ BOLTZMANN's constant $= 1\cdot37 \times 10^{-23}$ joules/°K.

$T_e =$ absolute temperature

$R =$ resistance of the conductor in ohms.

For the purpose of calculation it is convenient to consider the conductor as equivalent to a noise source generating a mean square

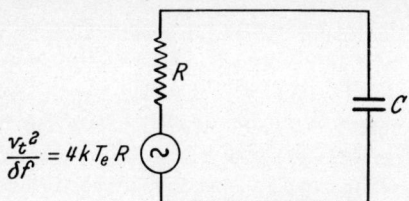

Fig. 12. Equivalent input circuit for thermal noise analysis

voltage per unit bandwidth of magnitude $4kT_eR$ and acting in series with a noiseless resistor of magnitude R. This follows directly from THEVININ's Theorem and is shown in fig. 11.

If the resistor R is shunted by a condenser C, as shown in fig. 12, the spectral density of the noise voltage appearing across the condenser is given by

$$\frac{v_t^2}{\delta f} = \frac{4kT_eR}{1 + (\omega CR)^2}. \qquad (6)$$

The total mean square noise amplitude across C effective throughout the complete frequency spectrum is obtained by integrating equation (6) from 0 to ∞. This gives

$$v_t^2 = 4kT_eR \int_0^\infty \frac{df}{1 + (\omega CR)^2}$$

Replacing df by $\dfrac{d\omega}{2\pi}$ and performing the integration gives

$$v_t^2 = \frac{4kT_eR}{2\pi CR}\left[\tan^{-1}(\omega CR)\right]_0^\infty$$

$$= \frac{kT_e}{C} \qquad (7)$$

Equation (7) is interesting because it shows that in any CR circuit the total thermal noise appearing across C is constant and equal to kT_e/C, and is also independent of the magnitude of R.

Curve (a) in fig. 13 shows how the spectral density of the noise across C varies with frequency.

For frequencies well below $f = 1/2\pi CR$ the intensity is constant and equal to $4kT_eR$. When $f = 1/2\pi CR$ the intensity has fallen to one half of its original value and for frequencies well above $1/2\pi CR$ the intensity drops rapidly as $1/f^2$. The total area under the curve is kT_e/C from equation (7). If the noise across C is amplified by an amplifier having a finite bandwidth extending from frequency f_1 to frequency f_2, then the area under the curve (a) and bounded by the f_1 and f_2 ordinates is a measure of the thermal noise passed by the amplifier. Curve (b) shows the effect of increasing the resistor by a factor 4. The low frequency intensity increases by the same factor but this time falls to one half at $f = 1/4(2\pi CR)$. Thereafter curve (b) quickly drops below curve (a), as it must do, since the total area under curve (b) has still to equal kT_e/C. It is clear then, that as far as the limited bandwidth amplifier is concerned, the total noise passed by it can be reduced by increasing R. In effect, an increase in R does not alter the total noise in the complete spectrum, but simply concentrates the noise energy in the lower frequencies of the spectrum and effects a reduction in the noise passed by an amplifier having a restricted low frequency cut off.

The input circuit of an ionisation chamber amplifier consists of the chamber capacity ΣC shunting the feed resistor R and the latter generates thermal noise in accordance with the above relation-

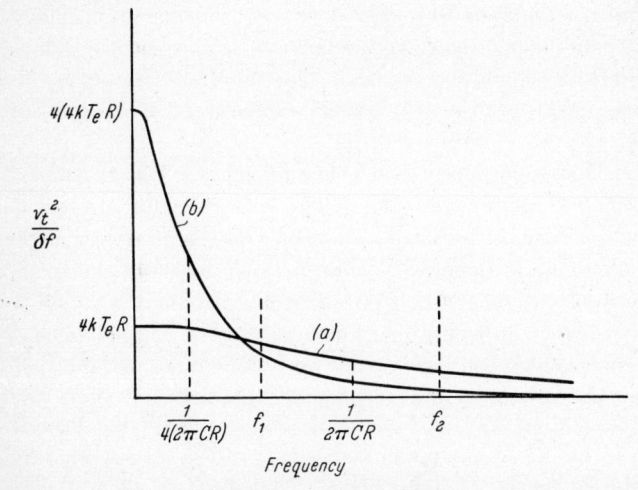

Fig. 13. Spectral density of thermal noise across a parallel RC circuit for two values of R

ships. If R is made large and the low frequency limit of the amplifier restricted by using a separate differentiating time constant, this case is similar to that just described and the thermal noise effective within the pass band of the amplifier depends on R and in particular decreases as R is increased. At the same time, making R large ensures that the signal charge from the chamber produces its full voltage $N_p e/\Sigma C$ across the chamber capacity, so the signal to thermal noise ratio clearly gets better as R is increased. In addition, by placing the differentiating time constant at some stage within the amplifier, any further sources of noise having a concentration of energy in the low frequency region of the spectrum (e.g. flicker noise, see Section 3.5) are limited in a similar way. The foregoing are the reasons, mentioned in Section 2.1, for making the input time constant

ΣCR very long compared with the chamber collection time, and using a separate differentiating time constant at some subsequent stage within the amplifier.

The thermal noise within the pass band of the amplifier can be decreased indefinitely by increasing R. There is, however, little point in continuing to increase R after the thermal noise has been reduced below other noise contributions yet to be considered, since excessively high values of R lead to difficulties of a more practical nature in the design of the amplifier circuit. In general therefore, thermal noise need never be considered in ionisation chamber amplifiers, except in so far as it is necessary to determine how large R should be, to keep the thermal noise insignificant. This point will be discussed further in Section 3.4, on grid current noise.

In the foregoing analysis on thermal noise it has been assumed that the noise relationships are valid from zero to infinite frequency. This is not strictly accurate, and the noise expressions no longer hold, when the frequency becomes comparable with the reciprocal of the time of flight of the free electrons between atomic collisions. This frequency, however, is of the order of a million megacycles and is well above any frequency of consequence within the pass band of the amplifiers at present under discussion.

3.3. SHOT NOISE

Shot noise is the name applied to the fluctuations of the mean current of a thermionic valve caused by the random emission of the electrons from the cathode. When a diode or a triode valve is operated under temperature limited conditions—that is, when all the electrons emitted from the cathode are collected by the anode—the spectral density of the current fluctuations is defined by the equation[13]

$$\frac{i_s^2}{\delta f} = 2eI_a \qquad (8)$$

where e = charge on an electron = $1 \cdot 59 \times 10^{-19}$ coulombs

I_a = mean anode current in amperes.

When the valve is operated under space charge limited conditions —that is, when the cathode emission is many times greater than the current taken by the anode—the current fluctuations in the anode circuit are also appreciably reduced and equation (8) is modified to take account of this by the introduction of a space charge smoothing factor F^2 (F^2 always < 1) to give

$$\frac{i_s^2}{\delta f} = 2eI_a F^2. \qquad (9)$$

For low power receiving triodes F^2 is of the order of 0·1.

When a valve is space charge limited, and this is the normal condition for operation, the cathode is surrounded by an electron cloud or space charge which has a potential, negative with respect to that of the cathode. This gives rise to a retarding field in the vicinity of the cathode and only those electrons emitted from the cathode with sufficient velocity to overcome the retarding field and pass through the space charge, reach the anode. All other electrons with emission velocities below this critical value are returned to the cathode and eventually an equilibrium condition is established between the emission current and the anode current. In a triode valve, the potential applied to the control grid effectively influences the potential of the space charge and thereby controls the anode current of the valve.

The mechanism of noise reduction by the space charge is very complicated[14] and will not be explained in detail here. A simple physical picture can be appreciated, however, by considering an excess of electrons emitted from the cathode with sufficient velocity to penetrate the space charge, the excess of electrons being fundamentally a fluctuation of the mean value. While in the vicinity of the space charge this excess of electrons temporarily lowers the potential of the space charge, with the result that other electrons having emission velocities only sufficient to penetrate the space charge are returned to the cathode instead. The magnitude of the fluctuation reaching the anode is thus effectively reduced by the electrons returned to the cathode, and from this very simple analysis it is not difficult to appreciate how the space charge can give rise to a reduction in, or a smoothing out of the noise.

The space charge smoothing factor F^2 has been the object of much theoretical and experimental work and it has been found that for valves with oxide coated cathodes, and this covers all the low power receiving valves considered in the text, a very good approximation is given by

$$F^2 = \frac{0.12\ g_m}{I_a}$$

where g_m = mutual conductance in amperes per volt.

It is convenient in most calculations to consider the shot noise, not as a current fluctuation, but as an equivalent voltage fluctuation referred to the grid of the valve. This modifies equation (9) to the form

$$\frac{v_s^2}{\delta f} = \frac{2eI_a F^2}{g_m^2}.$$

If now the value $\dfrac{0.12\ g_m}{I_a}$ is substituted for F^2 this gives

$$\frac{v_s^2}{\delta f} = \frac{0.24e}{g_m},$$

and finally by a rearrangement of the constants this can be written as

$$\frac{v_s^2}{\delta f} = 4kT_e \frac{2.5}{g_m} \qquad (10)$$

where T_e is assumed to be equal to 290°K—that is, normal room temperature.

If equation (10) is compared with equation (5) it will be appreciated that the equivalent shot noise voltage referred to the grid of a triode valve is equal to the thermal noise voltage which would be produced by a resistor of magnitude $2.5/g_m$ in series with the grid and at an ambient temperature of 290°K. This is a very convenient form in which to use the shot noise equation since it allows a quick estimation of the magnitude of the shot noise to be made from a knowledge of the valve mutual conductance. It is in fact the accepted custom to specify the shot noise of a valve by giving its equivalent noise resistance.

When the total valve current divides between two collecting electrodes, such as the anode and screen grid of a pentode valve, a further source of noise is introduced, known as partition noise. Although the mean partition of the total current between anode and screen is determined by the geometry of the valve, the final destination of each electron which passes the space charge is a matter of chance and this randomness gives rise to additional current fluctuations on each of the divided components.

The spectral density of the current fluctuations in the anode of a pentode valve is given by[14]

$$\frac{i_s^2}{\delta f} = 2eI_a \frac{F^2 I_a + I_s}{I_c},$$

where I_a I_s and I_c refer to the mean anode, screen and cathode currents respectively, and F^2 refers as before to the space charge smoothing factor of the cathode current fluctuations. If this expression is rearranged and put in the form of equation (10) above the result is

$$\frac{v_s^2}{\delta f} = 4kT_e \left[\frac{2 \cdot 5}{g_m} \frac{I_a}{I_c} \right] \left[1 + 8 \frac{I_s}{g_m} \right], \quad (11)$$

where g_m this time refers to the anode mutual conductance of the pentode valve. The expression in the first bracket, namely $\left[\dfrac{2 \cdot 5}{g_m} \dfrac{I_a}{I_c} \right]$ represents the equivalent noise resistance of a pentode valve when used as a triode under the same conditions of cathode current. Thus $\left[1 + 8 \dfrac{I_s}{g_m} \right]$ simply represents the number of times the equivalent noise resistance is increased by using the valve as a pentode.

Equation (10) defines the spectral density of the shot noise of a triode valve, expressed as an equivalent noise voltage at the grid. In the equivalent circuit of the ionisation chamber amplifier, shown in fig. 7, it is assumed that the grid of the first valve is connected to the point A. Hence equation (10) also defines the shot noise effective at the amplifier input terminals AA'. It is now required to calculate the magnitude of the shot noise at BB' in the equivalent circuit.

The spectral density at BB' allowing for the effect of the differentiating and integrating time constants T_1 and T_2 respectively is given by

$$\frac{v_s^2}{\delta f} = 4kT_e \frac{2 \cdot 5}{g_m} \frac{\omega^2 T_1^2}{1 + \omega^2 T_1^2} \frac{1}{1 + \omega^2 T_2^2}$$

where $\left[\dfrac{1 + \omega^2 T_1^2}{\omega^2 T_1^2} \right]^{1/2}$ is the attenuation of T_1 to sinusoidal voltage components and $[1 + \omega^2 T_2^2]^{1/2}$ is the attenuation of T_2 to sinusoidal voltage components. Hence the total mean square noise voltage at BB' is

$$v_s^2 = 4kT_e \frac{2 \cdot 5}{g_m} \int_0^\infty \frac{\omega^2 T_1^2 \, df}{(1 + \omega^2 T_1^2)(1 + \omega^2 T_2^2)}$$

$$= 4kT_e \frac{2 \cdot 5}{g_m} \int_0^\infty \frac{\omega^2 T_1^2 \, d\omega}{2\pi(1 + \omega^2 T_1^2)(1 + \omega^2 T_2^2)}$$

replacing df by $\dfrac{d\omega}{2\pi}$.

The integral is evaluated in Appendix C and is $\dfrac{T_1}{4T_2(T_1 + T_2)}$.

Therefore the total mean square shot noise at BB' in the equivalent circuit is

$$v_s^2 = kT_e \frac{2 \cdot 5}{g_m} \frac{T_1}{T_2(T_1 + T_2)}$$

and the r.m.s. voltage is

$$v_s = \left[kT_e \frac{2 \cdot 5}{g_m} \frac{T_1}{T_2(T_1 + T_2)} \right]^{1/2} \qquad (12)$$

From equation (12) it is clear that high slope valves must be used where low shot noise is required. Again, if one considers the simple case of an amplifier having a constant ratio between the differentiating and integrating time constants, i.e. $T_1/T_2 =$ constant, the shot noise is seen to vary as $\left[\dfrac{1}{T_1 + T_2}\right]^{1/2}$. Thus as T_1 and T_2 are decreased, corresponding to the amplifier pass band moving higher in the frequency spectrum, the shot noise will increase. In general therefore, high frequency amplifiers, used because of their good resolving power (see equation (4)), have a larger shot noise contribution than low frequency amplifiers having poor resolution.

Shot noise in valves has been investigated by other writers in great detail and good agreement between theory and experimental results can usually be expected. The author has made some experimental measurements, to check in particular equation (12) above but as these measurements also involve the flicker noise of the valve, it is proposed to present the results later in Section 3.5 after the characteristics of flicker noise have been described.

In the foregoing analysis on shot noise it has been assumed that the noise relationships are valid from zero to infinite frequency. Like thermal noise this is not the case and the noise expressions are no longer valid when the frequency becomes comparable with the reciprocal of the electron transit time in the valve. This frequency however is of the order of a thousand megacycles and is, as before, well above any frequency of consequence in the present discussion.

3.4 GRID CURRENT NOISE

In a thermionic valve, current can flow in the grid circuit for the following reasons,

(a) If the grid potential is not sufficiently negative with respect to the cathode, electrons passing through the space charge and on their way to the anode, may be captured by the grid. This effect obviously decreases as the negative grid potential is increased.

(b) Electrons comprising the anode current may collide with molecules of the residual gas left in the valve giving rise to the production of positive ions. The positive ions drift towards the cathode but are largely collected by the grid as this is the most negative electrode in the valve. Grid current due to positive ions flows in the opposite direction to that due to electrons and is usually referred to as reverse grid current. This effect decreases with the anode current of the valve.

(c) In valves having a low grid cathode spacing the grid may attain a sufficiently high temperature to cause electron emission. This gives rise to another component of reverse grid current which depends largely on the cathode temperature.

(d) Finally the cathode may emit positive ions which find their way to the grid and contribute a third component to the reverse grid current. This, too, is largely dependent on the cathode temperature.

There are other causes of grid current in thermionic valves, but all are of a sufficiently low order compared with those mentioned above to be completely neglected. In fact, for the valves which are considered in the text the main causes giving rise to grid current are (a) and (b), with causes (c) and (d) resulting only in second order contributions.

The general shapes of the grid current components of a low power receiving valve, plotted as a function of the negative grid voltage are shown in fig. 14.

The curve ab refers to the grid current component due to the capture of electrons from the main anode stream. This component

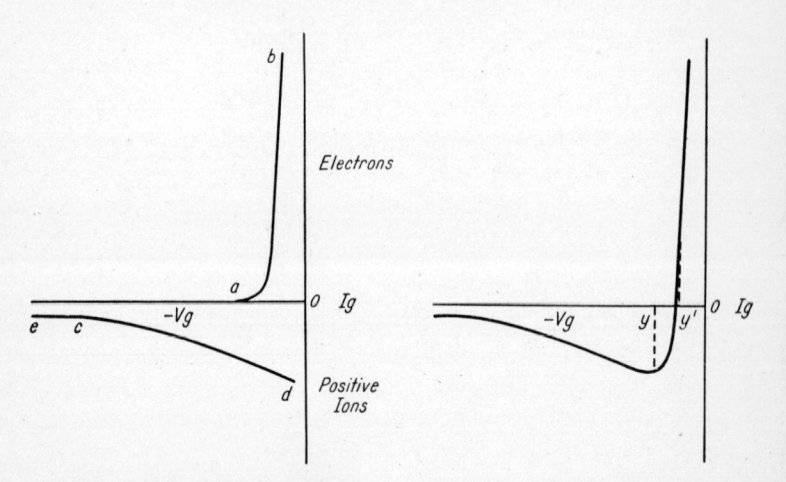

Fig. 14. Individual grid current components of low power receiving valve

Fig. 15. Composite grid current curve as obtained by experimental measurement

is usually insignificant for negative grid voltages in excess of about 2 volts but for negative grid voltages less than this figure, the electron grid current increases rapidly as shown.

Curve cd refers to the grid current component due to the production of positive ions in the residual gas. It decreases as the negative grid potential is increased since this corresponds to a reduction in the anode current and consequently a reduction in the number of ionising collisions.

Curve ce indicating a constant reverse grid current, refers to the combined effects of electron emission from the grid and positive ion emission from the cathode.

In fig. 14 the individual components of the grid current have been shown. In an experimental measurement of grid current however, the curve obtained would represent the algebraic sum of the grid current components. This is shown in fig. 15.

There is a small region yy' over which it is difficult to determine the individual components, but in view of the very rapid rise of the electron current component it is usually safe to assume that for negative grid potentials in excess of y the curve shows the true reverse grid current, while for potentials less than y' the curve can be taken as representing electron current.

Each component of the grid current has superimposed fluctuations due to the shot effect, and these current fluctuations in turn produce voltage fluctuations across the impedance in the grid circuit of the valve. This is known as grid current noise.

The spectral density of the grid current fluctuations is given by

$$\frac{i_g^2}{\delta f} = 2e(I_{g+} + I_{g-}) \qquad (13)$$

where I_{g+} = electron current in amperes
I_{g-} = reverse grid current in amperes.

Although the electron and reverse grid currents flow in opposite directions and therefore tend to cancel each other, the noise fluctuations due to each are completely random and consequently must be added. This is the reason for including the arithmetic sum of I_{g+} and I_{g-} in equation (13) above.

In view, however, of the very rapid rise in the electron current for low negative grid potentials, the noise in this region is not very stable and is very dependent on the grid potential. It is therefore always desirable to operate the valve in the region where the electron current is negligible and only the reverse grid current remains. It will hereafter be assumed that this condition is satisfied. Equation (13) may therefore be rewritten as

$$\frac{i_g^2}{\delta f} = 2eI_g$$

where I_g now refers to the reverse grid current.

In an amplifier for use with an ionisation chamber the impedance in the grid circuit comprises the total capacity ΣC in shunt with the feed resistor R. Hence the spectral density of the voltage fluctuations produced on the grid is given by

$$\frac{v_g^2}{\delta f} = 2eI_g \frac{R^2}{1 + \omega^2(\Sigma C)^2 R^2}. \qquad (14)$$

In section 3.2, on thermal noise, reasons were given for making the input time constant ΣCR large compared with the differentiating time constant in the amplifier. This means that $\omega \Sigma CR$ is much larger than unity for all frequencies within the pass band of the subsequent amplifier and thus equation (14) may be rewritten as

$$\frac{v_g^2}{\delta f} = \frac{2eI_g}{\omega^2 (\Sigma C)^2}$$

This represents the spectral density of the grid current voltage fluctuations at the input terminals AA' in the equivalent amplifier circuit. The spectral density at BB' allowing for the effects of the differentiating and integrating time constants T_1 and T_2 respectively is

$$\frac{v_g^2}{\delta f} = \frac{2eI_g}{\omega^2 (\Sigma C)^2} \frac{\omega^2 T_1^2}{1 + \omega^2 T_1^2} \frac{1}{1 + \omega^2 T_2^2}.$$

Hence the total mean square noise voltage at BB' is

$$v_g^2 = \frac{2eI_g}{(\Sigma C)^2} \int_0^\infty \frac{T_1^2 \, df}{(1 + \omega^2 T_1^2)(1 + \omega^2 T_2^2)}$$

$$= \frac{2eI_g}{(\Sigma C)^2} \int_0^\infty \frac{T_1^2 \, d\omega}{2\pi (1 + \omega^2 T_1^2)(1 + \omega^2 T_2^2)}$$

replacing df by $\dfrac{d\omega}{2\pi}$.

The integral is evaluated in Appendix D and is $\dfrac{T_1^2}{4(T_1 + T_2)}$.

Therefore, the total mean square grid current noise voltage at BB' is

$$v_g^2 = \frac{eI_g}{2(\Sigma C)^2} \frac{T_1^2}{T_1 + T_2}$$

and the r.m.s. voltage is

$$v_g = \left[\frac{eI_g}{2(\Sigma C)^2} \frac{T_1^2}{T_1 + T_2} \right]^{1/2}. \qquad (15)$$

From equation (15) it is clear that valves having a low order of grid current must be used if low grid current noise is required. Again, for the simple amplifier having a constant T_1/T_2 ratio, equation (15) shows that the grid current noise varies as $[T_1]^{1/2}$. Thus, as T_1 and T_2 are increased, corresponding to the pass band of the amplifier moving lower in the frequency spectrum, the grid current noise will likewise increase. In general, therefore, grid current noise is more troublesome in low frequency amplifiers in contradistinction to shot noise which tends to preponderate in high frequency amplifiers. In a practical amplifier it is usually necessary to consider both forms of noise since it will be shown later that the grid current noise can be improved at the expense of the shot noise and vice versa.

Since the grid current is due mainly to positive ions it depends largely on the amount of residual gas left in the valve. As a consequence the grid current can be expected to vary widely from one type of valve to another and also between valves of the same type. A large number of measurements have been made by the author on valves of different types to determine those which exhibit low orders of grid current. In general, it is found that the best results are obtained from low power medium slope receiving valves. Even so, the best valves differ largely among themselves and to achieve the lowest grid current possible, selection is usually found to be necessary.

In a selected valve the reverse grid current can be further reduced by

(a) Working the valve with increased negative grid potential. This reduces the anode current and with it the number of ionising collisions. For this reason one might expect the positive ion curve *cd* in fig. 15 to be proportional to the anode current over a limited range and this is found to be the case.

(b) Working the valve with decreased anode voltage (screen in the case of a pentode). This reduces the average kinetic energy of the ionising electrons and in consequence reduces the number of ionising collisions. It is to be expected, and is found in practice, that the positive ion current varies as the square of the anode voltage.

(c) Working the valve with reduced heater dissipation to keep grid emission of electrons and cathode emission of positive ions as low as possible.

Methods (a) and (b) improve the grid current at the expense of the mutual conductance. Thus the shot noise increases and the processes can not be continued indefinitely. In method (b) for instance, it has never been found possible to reduce the anode voltage to a level below the ionising potential of the residual gas. Shot noise always imposes a limitation before this stage is reached.

Method (c) also can not be carried to the extreme since this reduces the normal electron emission from the cathode, destroys the space charge and thereby causes the shot noise to increase. In any case, electron emission from the grid and positive ion emission from the cathode are of secondary importance and method (c) is only used on the very best of valves.

TABLE 1

Valve No.	$I_g \times 10^{-12} A.$	Valve No.	$I_g \times 10^{-12} A.$
1	50	11	9
2	6	12	> 100
3	7·5	13	10
4	27·5	14	12
5	> 100	15	> 100
6	15	16	12
7	37·5	17	7·5
8	5	18	20
9	> 100	19	> 100
10	7·5	20	40

Because of the very large variations which can occur in the grid current noise of valves it is always necessary to select the input valve of an ionisation chamber amplifier if grid current noise is at all important. During selection tests at A.E.R.E. it was found that no reliable figures for grid current could be obtained until the valves had been suitably aged. In general, the grid current of a new valve decreases with running time and may take anything up to two days to settle down. Once aged, a particular valve usually reaches its steady grid current value in about 15 to 30 min. after switching on from cold. At A.E.R.E. it is customary to age all valves for 48 hours under normal running conditions prior to making any grid current measurements.

Table 1 shows the results of spot measurements of grid current made on 20 aged valves type E.F.37 running under the conditions of $V_a = V_s = 45$ volts, $I_a = 150 \mu A$ and $V_h = 4\cdot 5$ volts.

The results show the wide variations which can occur among valves of a particular type and for a maximum grid current limit

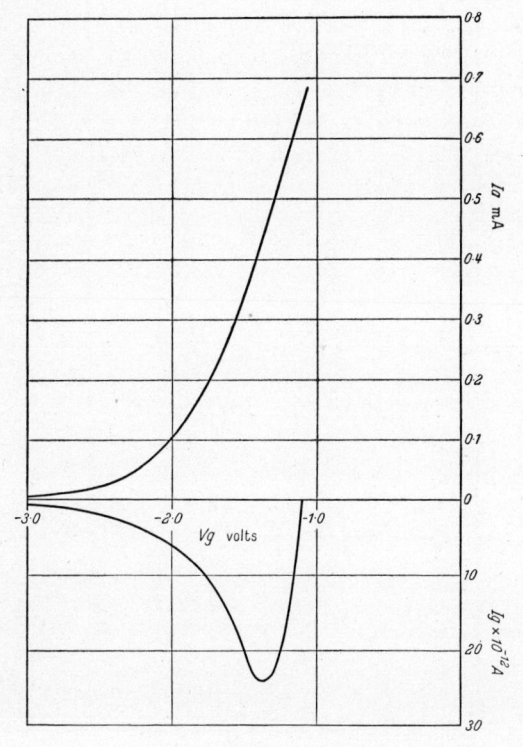

Fig. 16. Experimental anode current and grid current characteristics of ME 1400
Triode connected $V_a = 45$ volts
$V_h = 4\cdot 5$ volts

of 8×10^{-12} amp., valve numbers 2, 3, 8, 10 and 17 are acceptable and the rest are rejects. As a result of extensive selection of valves type EF37, for low grid current, it has been found that on the average one valve in every six is a good one. Approximately the same average figure holds for valves of other types.

The EF37 was found to be the best British valve exhibiting very low grid current and it has been used extensively in pulse amplifiers at A.E.R.E. So much so, that a quantity of several thousand valves were selected by the manufacturer for use at A.E.R.E. alone and it is now possible to buy such low grid current valves under the type number ME1400.

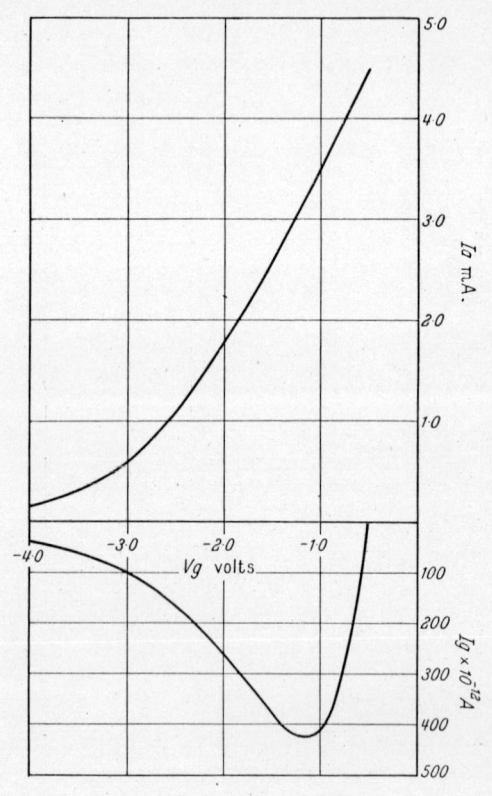

Fig. 17. Experimental anode current and grid current characteristics of ME 1400
Triode connected $V_a = 90$ volts
$V_h = 4·5$ volts

Fig. 16 shows experimental anode current and grid current characteristics of the ME1400 running as a triode with 45 volts on the anode and 4·5 volts on the heater. Suitable operating conditions for this valve are $I_a = 150 \mu A$, $g_m = 400 \mu A/\text{volt}$ and $I_g = 8 \times 10^{-12}$ amp.

Fig. 17 shows the experimental characteristics of the same valve running with an anode voltage of 90 volts. In this case, suitable operating conditions are $I_a = 1\text{mA}$, $g_m = 1\text{ mA/volt}$ and $I_g = 150 \times 10^{-12}$ amp.

The ME1400 is however a comparatively low slope valve, and although it gives excellent results in low frequency amplifiers

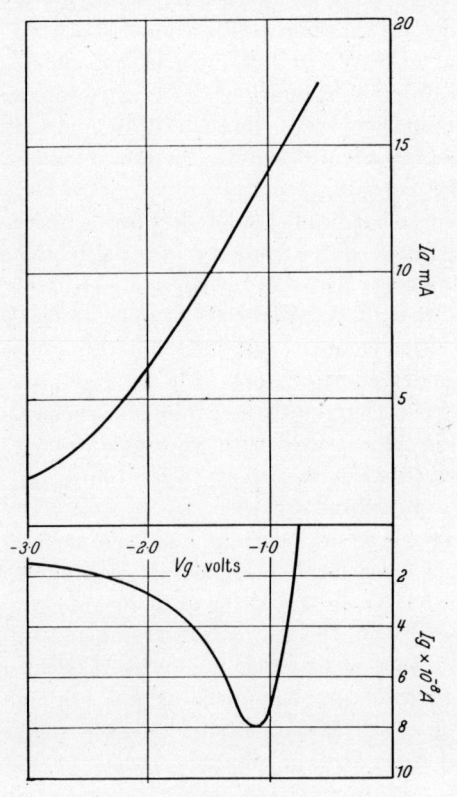

Fig. 18. Experimental anode current and grid current characteristics of selected EC 91
$V_a = 250$ volts $\qquad V_h = 6\cdot3$ volts

covering the range of a few hundred cycles per second to about 50 kc/s, it is not entirely suitable for use in high frequency amplifiers operating over the frequency range 50 kc/s to 5 Mc/s. At the very high frequencies in particular (order of 1 to 5 Mc/s) it is necessary to use a high slope valve, preferably a triode, to get minimum shot noise and so far in Britain, the EC91, selected for low grid current, has been largely used. This valve is a miniature, high slope triode and experimental anode current and grid current characteristics are shown in fig. 18 for an anode voltage of 250 volts. Suitable operating conditions are $I_a = 10\text{mA}$, $g_m = 8\cdot5\text{mA/volt}$, and $I_g = 5 \times 10^{-8}$ amp.

In America, the most commonly used valve is the 6AK5 pentode. This valve, when triode connected, has a slope slightly less than that of the EC91 but the grid current of selected valves is nearly an order of magnitude better. In addition, the 6AK5 is a lower dissipation valve than the EC91 and this is a factor of some consequence when designing the power supply to run a high gain amplifier containing many valves. The use of the 6AK5 valve in quantity has been restricted in this country, due to a very inadequate supply, but a number of British manufacturers are now producing equivalents and it is certain that this type of valve will soon be used to an appreciable extent, not only as the input valve of ionisation chamber amplifiers but also as a general purpose pentode for high frequency work. Fig. 19 shows the anode current and grid current characteristics, from experiment, of a selected 6AK5 running as a triode with an anode voltage of 120 volts. Suitable operating conditions are $I_a = 10\text{mA}$, $g_m = 6\cdot6\text{mA/volt}$ and $I_g = 7 \times 10^{-9}$ amp.

For a valve whose grid current has been accurately measured, the theoretical noise level is usually in good agreement with the measured noise level, provided the grid current refers to an operating point well within the ion current region and not to some point where electron current may also be flowing. In this latter case the measured noise level can be appreciably greater than that predicted from equation (15). Curves are included in Chapter 5 showing the measured noise level of amplifiers using the above mentioned input valves and from these curves the reasonable agreement between theoretical and experimental grid current noise can be observed.

In the foregoing section on grid current noise it has been assumed that the noise relationships are valid over the frequency range from zero to infinity. Like thermal and shot noise this is again not true and the expressions only hold up to frequencies comparable with the reciprocal of the transit time of positive ions in the valve. The transit time of a positive ion is appreciably longer than the transit time of an electron, and consequently the upper frequency of validity is here only of the order of tens of megacycles. Even so, this is sufficiently high for the noise relationships to be applicable especially when it is remembered that grid current is chiefly troublesome in low frequency amplifiers and is rarely of any consequence above 1 Mc/s.

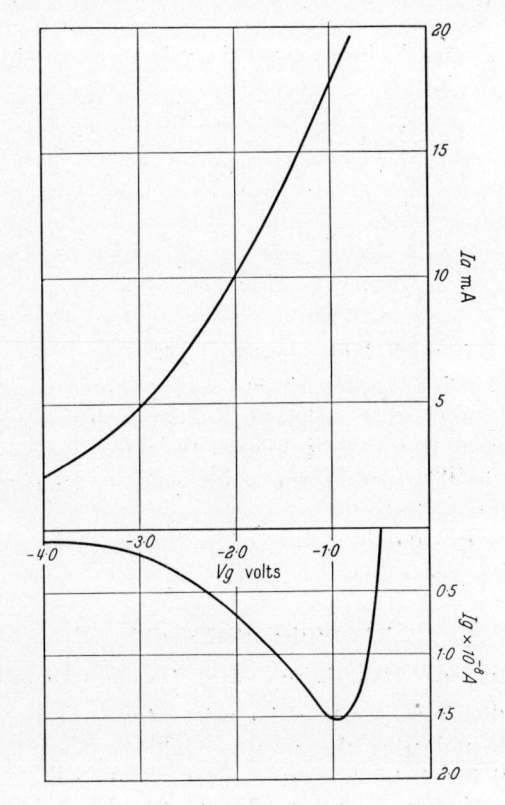

Fig. 19. Experimental anode current and grid current characteristics of selected 6AK5

Triode connected $V_a = 120$ volts
$V_h = 6\cdot 3$ volts

To conclude this section on grid current noise it only remains to effect the comparison mentioned in Section 3.2 between thermal noise and grid current noise. Inspection of equation (6) which defines the thermal noise spectral density at the amplifier input terminals AA' and of equation (14) which defines the grid current noise spectral density at the same point, shows that the two equations are identical in their frequency dependence and differ only in their constant terms. Thus the total thermal noise always bears a fixed ratio to the total grid current noise, irrespective of the amplifier bandwidth and for a given valve this ratio depends on the value of R.

For the thermal noise to equal the grid current noise equations (6) and (14) give

$$4kT_e R = 2eI_g R^2$$

$$\therefore R = \frac{4kT_e}{2eI_g}$$

$$\doteqdot \frac{0 \cdot 05}{I_g}. \qquad (16)$$

Thus if R is made large compared with $0 \cdot 05/I_g$ (say ten times) the thermal noise is always insignificant compared with the grid current noise and can be neglected. For the two operating conditions of the ME1400, for the EC91 and for the 6AK5 valves previously mentioned in this section, the shunt resistors required to make the thermal noise insignificant in each case are 6×10^{10}, $3 \cdot 5 \times 10^9$, 10^7 and 7×10^7 ohms respectively.

3.5. FLICKER NOISE

Flicker noise in valves with oxide coated cathodes is due[15] to the random appearance of impurity centres on the surface of the cathode, affecting the electron emission over the impurity areas and giving rise to fluctuations of the anode current. Although the exact mechanism whereby the impurity centres appear on and disappear from the cathode surface is not fully understood, the characteristics of flicker noise are fairly well known from experimental measurements. It has been shown[16]

(a) That the flicker noise voltage over a given bandwidth referred to the grid, does not usually vary outside $\pm 50\%$ limits on a constant value for most types of receiving valves.

(b) That the flicker noise voltage referred to the grid is sensibly independent of the anode current, provided the valve is running under space charge limited conditions.

(c) That the spectral density of the flicker noise referred to the grid varies as $1/f$ over a considerable portion of the frequency spectrum.

The spectral density of flicker noise at the grid of a valve is approximately given by[16]

$$\frac{v_f^2}{\delta f} = \frac{10^{-13}}{f} \qquad (17)$$

where the factor 10^{-13} is an average figure obtained from experimental measurements and the variation as $1/f$ has been checked experimentally to hold for frequencies from very much less than 1 c/s to at least 50 kc/s. It will be shown later in this section that in pulse amplifiers for use with ionisation chambers, the flicker noise is always insignificant compared with the shot noise above 50 kc/s and therefore equation (17) can be taken as valid over the whole frequency range of present interest.

The spectral density of the flicker noise at BB' in the equivalent amplifier circuit, allowing for the effects of the differentiating and integrating time constants T_1 and T_2 respectively is

$$\frac{v_f^2}{\delta f} = \frac{10^{-13}}{f} \frac{\omega^2 T_1^2}{1 + \omega^2 T_1^2} \frac{1}{1 + \omega^2 T_2^2}.$$

Hence the total mean square noise voltage at BB' is

$$v_f^2 = 10^{-13} \int_0^\infty \frac{\omega^2 T_1^2 \, df}{f(1 + \omega^2 T_1^2)(1 + \omega^2 T_2^2)}$$

$$= 10^{-13} \int_0^\infty \frac{\omega T_1^2 \, d\omega}{(1 + \omega^2 T_1^2)(1 + \omega^2 T_2^2)}$$

replacing $\dfrac{df}{f}$ by $\dfrac{d\omega}{\omega}$.

The integral is evaluated in Appendix E and is

$$\frac{T_1^2}{T_1^2 - T_2^2} \log_\varepsilon \frac{T_1}{T_2}.$$

Therefore the total mean square flicker noise voltage at BB' in the equivalent circuit is

$$v_f^2 = \frac{10^{-13} T_1^2}{T_1^2 - T_2^2} \log_\varepsilon \frac{T_1}{T_2},$$

and the r.m.s. voltage is

$$v_f = \left[\frac{10^{-13} T_1^2}{T_1^2 - T_2^2} \log_\varepsilon \frac{T_1}{T_2} \right]^{1/2}. \tag{18}$$

For the simple amplifier having a constant T_1/T_2 ratio equation (18) shows that the flicker noise is also constant. Thus, as T_1 and T_2 are increased or decreased, corresponding to the pass band of the amplifier moving higher or lower in the frequency spectrum the flicker noise will not vary. This should be compared with shot noise which increases in high frequency amplifiers and grid current noise which increases in low frequency amplifiers.

Consider the special case of an amplifier having equal differentiating and integrating time constants, i.e. $T_1 = T_2$. The r.m.s. flicker noise at BB' in the equivalent circuit of this amplifier is given by equation (18) above. The right hand side of equation (18) has a limiting value when $T_1 = T_2$ and this is shown in Appendix F to be $\left[\dfrac{10^{-13}}{2} \right]^{1/2} = 0 \cdot 22 \mu\text{V}$. This value for the r.m.s. flicker noise is constant for all values of T_1 within the valid range of equation (18). The shot noise for the same amplifier is obtained from equation (12) by making $T_1 = T_2$ and is

$$\left[kT_e \frac{2 \cdot 5}{g_m} \frac{1}{2T_1} \right]^{1/2} = 7 \times 10^{-9} \left[\frac{1}{g_m T_1} \right]^{1/2} \text{volts.}$$

The shot noise is dependent on T_1 and in particular varies as $\left[\dfrac{1}{T_1} \right]^{1/2}$. Thus with a short circuit across the input, thereby reducing to zero any thermal or grid current noise components, a plot of the residual noise against T_1 for this amplifier should show a predominance of shot noise for low values of T_1, with a tendency to fall to a constant level representing the flicker noise, for high values of T_1. This relationship was verified in the following experiment.

The amplifier used had equal differentiating and integrating time constants, adjustable together over a wide range. The input was short circuited with a large condenser and the noise at the amplifier output was measured on a thermocouple instrument calibrated to read r.m.s. volts. To convert the output noise figures to the corresponding values at BB' in the equivalent circuit, measurements of the amplifier gain G were also made. These latter measurements were carried out by two different methods, designed to give a cross check on the experimental figures. In method (1) a sinusoidal voltage of known amplitude and of frequency centred on the middle of the amplifier pass band, i.e. $1/2\pi T_1$, was fed to the input and the output voltage measured. From these figures the gain G was computed allowing for an attenuation of 2 suffered by the sinusoidal voltage in passing through the equal time constants T_1 and T_2. Method (2) involved the use of pulses. A rectangular pulse, having a very fast rising edge, a very long duration and of known amplitude was fed to the amplifier input and the amplitude of the output pulse measured. Again, from the input and output figures the gain G was calculated, this time allowing for a pulse attenuation of approximately $1/0.37$ by the equal time constants T_1 and T_2 (see curves in fig. 9). Gain checks were made for every value of T_1 and T_2 to compensate for possible inequality due to component tolerances and the final gain figures used in the experiment were the averages of each pair of values obtained by the two different methods. The input valve used was the EC91 running under the conditions $V_a = 250$ volts $I_a = 10$ mA and $g_m = 8.5$ mA/volt. The experiment was repeated for ten different input valves and the figures given in Table 2 below show the average results.

TABLE 2

$T_1 = T_2$	f	Gain G	Experimental Shot + Flicker Noise	Theoretical Shot + Flicker Noise
μs	kc/s	$\times 10^6$	μV	μV
3·2	50	4·88	0·48	0·485
8·0	20	5·78	0·38	0·35
16	10	6·63	0·30	0·29
32	5	6·69	0·25	0·26
80	2	6·28	0·21	0·23
160	1	6·99	0·195	0·22
320	0·5	6·85	0·195	0·22

The experimental figures show clearly the expected drop in the noise with increasing T_1 and T_2 and the tendency to reach a constant value representing the flicker noise. The experimental value for the flicker noise is 0·195 μV which is some 13% below that given by equation (18). Of the ten valves tested, the lowest flicker noise measured was 0·14 μV and the highest was 0·26 μV. The former is 35% low on the figure from equation (18) and the latter 20% high. The results in the table are in good agreement with previous measurements and would appear to support the experimental characteristics of flicker noise listed at the beginning of this section.

For the shortest time constant values only the shot noise is significant, and here again agreement between theory and experiment is very good. For the EC91 and for the special bandwidth defined by $T_1 = T_2$ the flicker noise is obviously of no consequence above 50 kc/s. This also holds for most other valves, as mutual conductances in excess of 8·5 mA/volt are improbable and therefore the shot noise figures in Table 2 can be taken as representing minimum values. Again, it will be shown in Section 4.3 that the type of bandwidth used in this experiment is representative of ionisation chamber amplifiers and thus equation (18) gives a reasonable measure of the flicker noise throughout the frequency range of immediate interest.

3.6. OTHER FORMS OF NOISE

For the sake of completeness the following additional sources of noise are described. In every case however it will be shown that the extra noise contributions can be removed entirely or are negligible in comparison with the contributions from the main sources detailed in Sections 3.2 to 3.5 inclusive.

3.6.1. Induced Grid Noise.

This type of noise arises from the induction effect of the electrons in the main anode stream, on the grid of a thermionic valve. As an electron approaches the grid a pulse of current flows in the grid circuit in one direction and as the electron moves away from the grid a pulse of current flows in the opposite direction. The time integrals of the two current pulses are equal and opposite and no resultant D.C. effect is measurable in the grid circuit. In a very idealised case the current pulses can be represented by the waveform in fig. 20.

The time τ between the pulses corresponds to the transit time of the electron and is of the order of 10^{-9} sec. The waveform in fig. 20 can be regarded as the sum of two current pulses, one positive, one negative, with the latter delayed a finite time τ with respect to the former. The induced noise current is of course caused by the random succession of a very large number of such pairs of pulses and it is the object of this analysis to now determine the form of the spectral density of this noise current. The spectral density of the positive current pulses is similar to that for shot

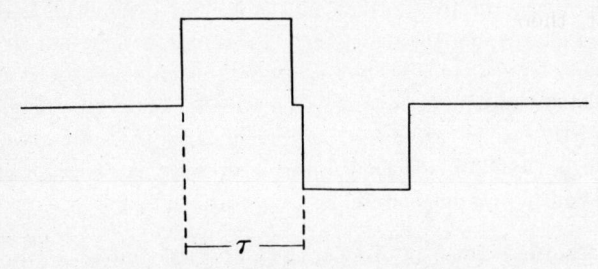

Fig. 20. Idealised waveform of induced grid current pulse

noise and therefore independent of frequency, and the same can be said of the spectral density of the negative current pulses. The summation, however, has a spectral density which varies as f^2 because similar frequency components of each pulse have a fixed time difference τ between them, and their summation accordingly varies as f, provided f is very much smaller than $1/\tau$. This is true for all frequencies under consideration in the text.

Thus, the spectral density of the induced grid current fluctuations is given by

$$\frac{i^2}{\delta f} = J_1 f^2$$

where J_1 is a constant depending on the anode current I_a and the transit time τ.

The voltage fluctuations produced across ΣC ($\omega \Sigma C R$ assumed large compared with unity) therefore have a spectral density given by

$$\frac{v^2}{\delta f} = \frac{J_1 f^2}{\omega^2 (\Sigma C)^2}. \tag{19}$$

In equation (19) the frequency terms cancel and the conclusion drawn is that the spectral density of the voltage fluctuations induced on the grid is similar to that for the shot noise fluctuations also referred to the grid. Consequently if induced noise is at all important it should show up as a contribution to the normal shot noise. In fact, there is almost certainly a correlation between the two types of noise, because the grid can be pictured as behaving in the same fashion as a space charge. An excess of electrons approaching the grid will lower its potential by the induction effect and thereby cause other electrons to return to the cathode. Thus, the contribution of the induced grid noise should tend to reduce the magnitude of the normal shot noise, in a manner analagous to space charge smoothing.

In fact, from extensive measurements of shot noise throughout the relevant frequency range, agreement between theory and experiment has always been very good, and the only conclusion possible is that induced grid noise, if present, is insignificant and can be completely neglected.

3.6.2. Thermal Noise in Anode Load of First Valve.

The anode resistor of the first valve generates thermal noise in accordance with equation (5) and it is of interest to compare this noise contribution with the normal shot noise of the valve.

Consider the simple triode amplifier shown in fig. 21.

R_a is the value of the anode resistor and g_m, ρ and μ refer to the mutual conductance, anode impedance and amplification factor respectively of the valve.

$$\text{Gain from grid to anode} = \frac{\mu R_a}{R_a + \rho}$$

$$= \frac{g_m \rho R_a}{R_a + \rho}.$$

The r.m.s. thermal noise voltage of R_a over a given bandwidth is proportional to $[R_a]^{1/2}$. Thus the fraction of this voltage appearing on the anode is proportional to $\dfrac{[R_a]^{1/2} \rho}{R_a + \rho}$ and the equivalent voltage referred to the grid is proportional to $\dfrac{[R_a]^{1/2}}{g_m R_a}$.

The shot noise voltage on the grid for the same bandwidth is proportional to $\left[\dfrac{2 \cdot 5}{g_m}\right]^{1/2}$.

Thus
$$\frac{\text{Shot Noise Voltage}}{\text{Thermal Noise Voltage}} = \left[\frac{2\cdot 5}{g_m}\right]^{1/2} \frac{g_m R_a}{[R_a]^{1/2}}$$
$$= [2\cdot 5\, g_m R_a]^{1/2}.$$

Here $g_m R_a$ is a factor which is considerably greater than unity if the stage is giving any worthwhile gain. Thus the thermal noise is small compared with the shot noise of the valve and can as a consequence be neglected.

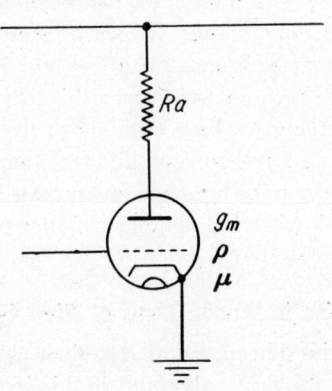

Fig. 21. Single stage triode amplifier

3.6.3. Current Noise in Semi-conductors.

In a resistor made of a semi-conducting material like carbon and carrying a current it is found that in addition to the normal thermal noise there is another component of noise which is a function of the magnitude of the current flowing in the resistor. This type of noise is due to fluctuations in the contact resistance between adjacent granules and has a spectral density given by

$$\frac{v^2}{\delta f} = \frac{J_2 I^2}{f}$$

where J_2 is a constant depending on the resistor material and I is the current flowing in the resistor.

The feed resistor to an ionisation chamber is always very large (see equation (16)) and the use of a carbon composition type of resistor is a necessity. Moreover, the coupling between the ionisation chamber and the amplifier is usually arranged in such a way that the feed resistor R serves as a grid leak for the input valve.

The feed resistor therefore carries the grid current of the valve and in consequence generates current noise.

The spectral density of the voltage fluctuations across ΣC from this source is given by

$$\frac{v^2}{\delta f} = \frac{J_2 I_g^{\,2}}{\omega^2 (\Sigma C)^2 f}. \tag{20}$$

Equation (20) varies as $1/f^3$ so this type of noise is largely concentrated in the low frequency region of the spectrum. The author has carried out a series of measurements on an ionisation chamber amplifier at a frequency of several hundred cycles per second and has not been able to detect this form of noise. Various values of feed resistor R were used, all above the critical value given by equation (16) and resistors of different make were also tried but in every case the noise was not measurable and it is concluded that for all normal operating conditions, current noise in the feed resistor R can be neglected.

3.6.4. Current Noise in Anode Load of First Valve.

The current noise phenomenon described above can occur in the anode load of the first valve and if this is a composition type resistor, can be significant on account of the much higher currents which are flowing. However, the resistor value here is never likely to exceed 10^5 ohms and in all probability will be very much lower, in which case a wire wound resistor can be used and there is then no possibility of current noise being produced. The author has always made a point of using a wire wound resistor for the anode load of the first valve and even for frequencies up to 10 Mc/s the inductance of the wire wound resistor has never given any trouble.

4

SIGNAL TO NOISE RATIO

4.1. GENERAL

The signal to noise ratio of an ionisation chamber and pulse amplifier is usually specified by comparing the maximum pulse signal at BB' in the equivalent circuit with the total noise at the same point.

In an experiment for the measurement of the energies of particles emitted from a radioactive source, it is customary to plot, by means

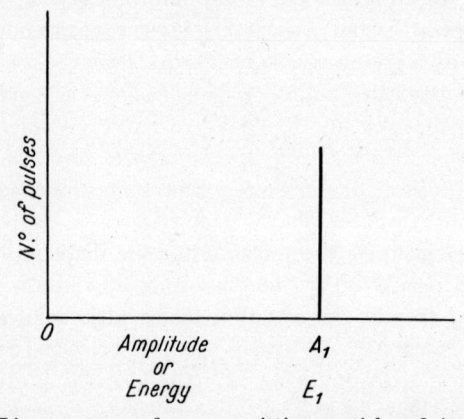

Fig. 22. Line spectrum of source emitting particles of single energy

of an instrument known as a kicksorter[17], a curve showing the amplitude distribution of the pulses at the amplifier output. This instrument consists of a large number of amplitude discriminators, each discriminator having its trigger level set to a slightly higher amplitude than its predecessor. The pulses from the amplifier are fed to all the discriminators simultaneously and the instrument automatically sorts the pulses into groups having amplitudes lying within the channels between adjacent discriminators and records the number of pulses in each group during the time of the experiment.

In the simple case of a source emitting particles of a single energy, and assuming for the moment an ideal noise free amplifier, and no other sources of fluctuations in the system, the resulting amplitude spectrum is as shown in fig. 22.

This consists of a single vertical line indicating that all the pulses have the same amplitude A_1 and is known as a line spectrum. If the relation between particle energy and pulse amplitude at the amplifier output is known from a previous calibration with a standard source, the amplitude spectrum in fig. 22 can also be regarded as representing the energy spectrum E_1 of the unknown source.

In a practical case, where the amplifier has a finite noise level and where, by careful design of the chamber[18] and preparation of the source, all other fluctuations in the system can usually be neglected in comparison, the curve given by the kicksorter is modified and has the form shown in fig. 23.

All the pulses no longer have an amplitude exactly equal to A_1. The majority have, but others have their amplitudes increased or decreased by varying amounts due to the noise. For the types of noise* discussed in Chapter 3 the contour of the spectrum peak at A_1 follows the well known Gaussian curve in which the height of an ordinate, displaced by an amount x from the maximum ordinate, varies as $\varepsilon^{-\frac{1}{2}(x/v)^2}$, where v represents the r.m.s. amplitude of the noise.

Since the contour of the pulse amplitude distribution about A_1 is determined primarily by the magnitude of the r.m.s. noise, it is convenient to specify the signal to noise ratio in experiments of

---

* The noise arises fundamentally from the random succession of a very large number of single events, [electrons reaching the anode for shot noise and positive ions reaching the grid for grid current noise.] Each event produces a pulse at BB' in the equivalent circuit having an effective duration Δ determined by the time constants T_1 and T_2. If $\bar{N}$ is the mean number of noise events per second then $\bar{N}\Delta$ is the mean number in the time of a single pulse. $\bar{N}\Delta$ is always an extremely large number. The probability of getting $N\Delta$ events in a time Δ is given by the POISSON formula and is $\varepsilon^{-\bar{N}\Delta}\dfrac{(\bar{N}\Delta)^{N\Delta}}{N\Delta!}$. This expression is unwieldy when $\bar{N}\Delta$ is very large but can be reduced to the formula of GAUSS, namely $\dfrac{1}{(2\pi\bar{N}\Delta)^{1/2}}\varepsilon^{-\frac{1}{2}\frac{(N\Delta-\bar{N}\Delta)^2}{\bar{N}\Delta}}$ which is equal to $\dfrac{1}{(2\pi\bar{N}\Delta)^{1/2}}\varepsilon^{-\frac{1}{2}\frac{D_e^2}{\bar{N}\Delta}}$ where $D_e = N\Delta - \bar{N}\Delta$ is the deviation from the mean value and $\bar{N}\Delta$ is of course also equal to the mean square deviation. The noise amplitude is linearly related to the fluctuation of the number of events occurring in time Δ and thus the probability of the noise reaching a value x where the r.m.s. value is v, is also given by the GAUSS formula and is $\dfrac{1}{(2\pi)^{1/2}v}\varepsilon^{-\frac{1}{2}(x/v)^2}$. This approach to random noise is discussed in more detail in Section 5.2.

SIGNAL TO NOISE RATIO

this nature by comparing the signal amplitude A_1 directly with the r.m.s. noise. This is a precise definition of signal to noise ratio

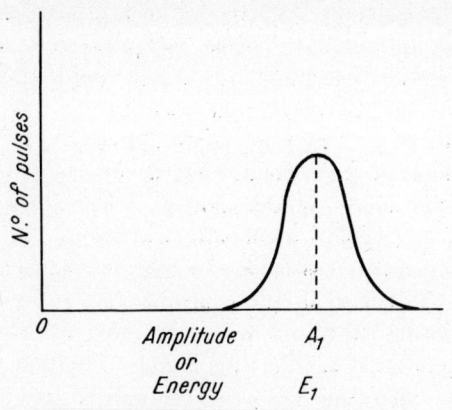

Fig. 23. Effect of noise on line spectrum

and is utilised to an appreciable extent in subsequent sections of the text.

In experiments on the other hand for the determination of the total activity of an unknown source, accurate measurements of

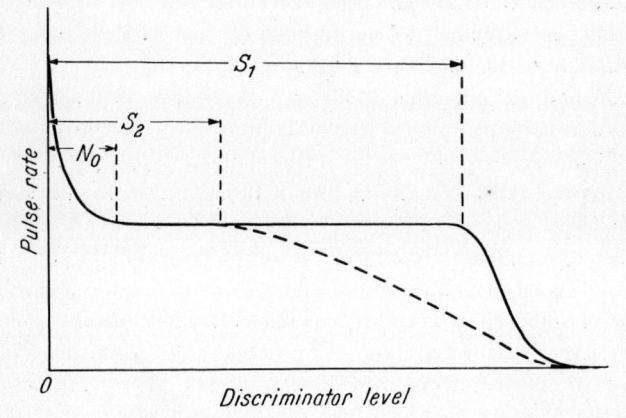

Fig. 24. Typical discriminator bias curve showing counting plateau

pulse rate are required. Here it is customary to use only a single amplitude discriminator and to adjust the trigger level to record as many signal pulses as possible and as few noise crests as possible. A typical curve relating the pulse rate to the discriminator level is shown in fig. 24.

The high pulse rate at low discriminator levels is due to the counting of noise crests. In the region above this, where the pulse rate is virtually constant, only the signal pulses are being counted. At still higher discriminator levels the pulse rate begins to fall until ultimately the discriminator level is greater than the maximum pulse amplitude and the count rate is zero.

The flat part of the curve is usually referred to as the counting plateau and under normal conditions the discriminator is set to a point about the middle of the plateau, which gives the greatest margin of safety against instrumental variations. In experimental work of this type it is customary to base the signal to noise ratio definition on the length of the counting plateau and for the full curve shown in fig. 24 this can be expressed by the ratio S_1/N_0. It will be appreciated that in this simple definition the points S_1 and N_0 on the curve are not precisely located and therefore this signal to noise ratio definition as used in such experiments is likewise not a precise one. Since the prime object is, however, to obtain a sufficiently long plateau to make the count rate reasonably immune from instrumental variations, large and precise signal to noise ratios are not necessary. A signal to noise ratio of 2 or 3 is usually adequate. In this case chambers for total activity measurements need not be designed to accommodate the whole track of the energised particle, and this simplifies enormously the geometrical and mechanical problems involved. A simple activity chamber might, for instance, give a curve as shown by the dotted line in fig. 24. The signal to noise ratio of this chamber is S_2/N_0 but provided this does not fall below about 2 or 3 the chamber should be quite satisfactory.

If for experimental reasons it becomes very difficult to achieve a signal to noise ratio in excess of 2, then in assessing the performance of the chamber it is necessary to take other factors into account, in particular the maximum permissible noise counting rate in relation to the counting rate of the signal. These considerations make the problem very involved and it is not proposed to discuss them further here. They will, however, be treated fully in Section 5.2 on the sensitivity of amplifiers used for total activity measurements.

In the majority of activity experiments, however, the simple signal to noise ratio definition given above is sufficient for all practical purposes.

4.2. SIGNAL TO NOISE RATIO IN TRIODES AND PENTODES

Consider a pentode valve used in the input stage of an ionisation chamber amplifier. The relevant circuit details are shown in fig. 25.

C is the chamber capacity, C_g and C_s are the grid to cathode and screen to grid capacities respectively of the valve and Q represents the total charge collected in the ionisation chamber.

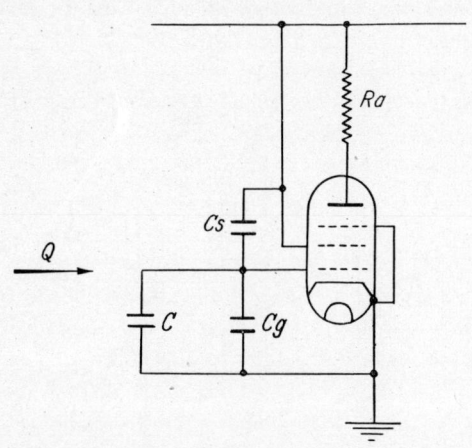

Fig. 25. Pentode input circuit for noise analysis

The voltage signal on the grid $= \dfrac{Q}{C + C_g + C_s}$

$$= \dfrac{Q}{\Sigma C}$$

where ΣC is the total cold input capacity.

If now v_{pen} represents the total r.m.s. shot noise voltage on the grid of the pentode, defined by equation (11) and taken over the pass band of the subsequent amplifier, the signal to shot noise ratio is given by

$$\dfrac{S}{N_{(\text{Shot})}} = \dfrac{Q}{\Sigma C v_{\text{pen}}}. \qquad (21)$$

Equation (21) shows that the signal to shot noise ratio can be improved by reducing ΣC, and this is clearly a general conclusion applicable to all types of input circuit. Reduction of ΣC also improves the signal to flicker noise ratio but does not alter the signal to grid current noise ratio since here, both signal and noise vary as $1/\Sigma C$.

If the pentode in fig. 25 is now connected as a triode, the circuit is shown in fig. 26.

The mutual conductance, amplification factor and anode impedance of the valve are represented by g_m, μ and ρ respectively, and C_a represents the anode to grid capacity (which is, of course, equal to C_s since the same valve is being used).

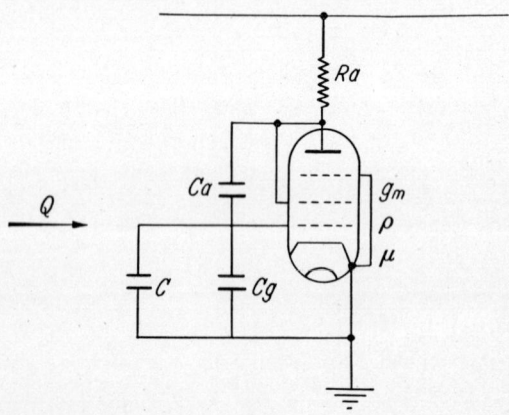

Fig. 26. Input circuit with valve connected as a triode

It is now required to determine whether the signal to shot noise ratio is affected by operating the valve as a triode. Let v_{tri} represent the total r.m.s. shot noise voltage on the grid of the triode, defined by equation (10) and taken over the pass band of the subsequent amplifier. The gain M of the stage to any voltage appearing between grid and cathode is given by

$$M = \frac{\mu R_a}{R_a + \rho}.$$

The input capacity allowing for the MILLER effect is

$$C + C_g + C_a(1 + M) = \Sigma C + MC_a.$$

Therefore the voltage signal between grid and cathode is $Q/(\Sigma C + MC_a)$ and the signal on the anode is given by

$$\text{Anode Signal} = \frac{QM}{\Sigma C + MC_a}. \tag{22}$$

As far as the shot noise is concerned, C_a in conjunction with C and C_g can be regarded as a negative feedback loop applying the fraction $C_a/\Sigma C$ of the anode voltage back to the grid. If, therefore, v_a represents the r.m.s. shot noise voltage actually appearing on the anode under the circuit conditions of fig. 26., the voltage fed back to the grid is $(v_a C_a)/\Sigma C$. The noise voltage required on the grid to produce v_a in the anode is v_a/M and this must represent the difference between the theoretical grid noise v_{tri} and the voltage $(v_a C_a)/\Sigma C$ fed back from the anode.

Hence

$$v_{tri} - \frac{v_a C_a}{\Sigma C} = \frac{v_a}{M}$$

giving

$$v_a = \frac{v_{tri} M \Sigma C}{\Sigma C + MC_a}. \qquad (23)$$

Therefore, the signal to shot noise ratio is

$$\frac{S}{N}_{(Shot)} = \frac{QM}{\Sigma C + MC_a} \frac{\Sigma C + MC_a}{v_{tri} M \Sigma C}$$

$$= \frac{Q}{\Sigma C \, v_{tri}}. \qquad (24)$$

Equation (24) is an extremely important one and leads to the following conclusions

(a) By using a triode instead of a pentode in the input stage of an ionisation chamber amplifier, the signal to shot noise ratio can be improved by the factor $\dfrac{\text{pentode noise } v_{pen}}{\text{triode noise } v_{tri}}$. That is, full advantage can be taken of the removal of partition noise. The improvement may be about 2 to 3 times for average receiving valves.

(b) Since C_a in conjunction with C and C_g effectively apply negative feedback to the triode when used under the circuit conditions of fig. 26, it can be concluded that negative feedback does not alter the signal to noise ratio in a circuit, provided the bandwidth of the subsequent amplifier is not changed and provided the source of noise is contained within the feedback loop. This is a perfectly general conclusion and is applicable to all types of feedback circuits and to all types of noise.

(c) In equation (24) the capacity appearing in the denominator is ΣC, the total cold input capacity, while v_{tri} represents the shot noise of the triode as defined by equation (10). Thus in all computations of signal to noise ratio, where the theoretical formulae developed in Chapter 3 are used, the signal for comparison must be obtained by initially considering the chamber charge Q as flowing into the total cold input capacity ΣC. This again is a perfectly general conclusion, applicable to all types of noise.

The use of a triode in the input stage of an ionisation chamber amplifier is therefore to be recommended on account of the resulting improvement in the signal to shot noise ratio, while the signal to grid current noise and the signal to flicker noise ratios remain unaltered.

The triode, as might be expected, is not without its disadvantages, but these are of a minor nature and only arise when the chamber capacity C is very small. Equation (23) shows that the noise on the triode anode is $\dfrac{v_{\text{tri}} M \Sigma C}{\Sigma C + MC_a}$ which can be rewritten in the form $v_{\text{tri}} \left[\dfrac{M}{1 + \dfrac{MC_a}{\Sigma C}} \right]$. The term in the brackets can be regarded as giving the effective gain of the triode since it defines the number of times the noise on the anode exceeds the theoretical grid noise v_{tri}. The effective gain is clearly dependent on the chamber capacity C, especially if MC_a is much larger than ΣC, a condition which will almost certainly arise when C is very small. This is an inconvenience largely of a practical nature, since it introduces difficulties into the design of an input circuit which can be used with a variety of chambers having different capacities. Again if MC_a is much larger than ΣC the effective gain of the triode is approximately $\Sigma C/C_a$. For very low chamber capacities this may be as small as 3, and then the noise of the following stage must be taken into consideration. If the following stage is a triode having an equally high slope its shot noise contribution will not be appreciable. If it is a pentode however it may be comparable, so a very low chamber capacity also imposes some conditions on the valve used in the second stage of the amplifier. This, too, may be a practical inconvenience since it is customary to use pentodes wherever possible in high gain amplifiers.

A circuit using two triodes, which is no better in signal to shot noise ratio for low chamber capacities, than the conventional arrangement just described but which has a more constant gain and gives certain practical advantages, is shown in fig. 27.

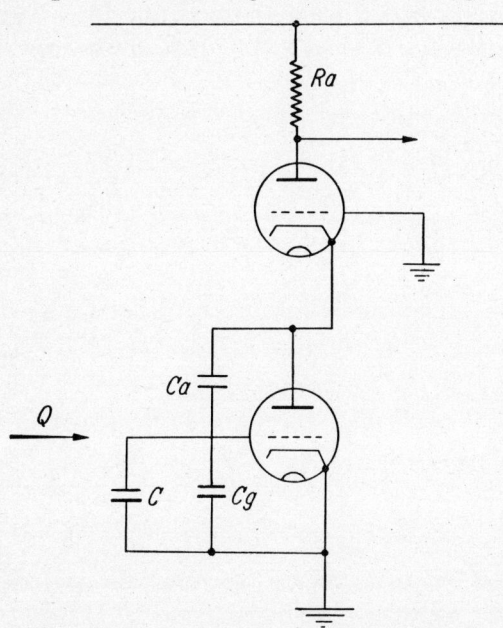

Fig. 27. Double triode input circuit—"cascode"

The anode load of the bottom valve is the impedance seen looking into the cathode of the top valve. This impedance is very low, so the gain to the anode of the bottom valve is likewise low. The MILLER effect on the bottom valve is thus not important and for this reason the overall gain of the two valve system is much less dependent on the chamber capacity C, than in the case of a single triode. A detailed analysis of this circuit is given in Appendix G and there it is shown that the overall effective gain is

$$\frac{g_m R_a}{1 + \frac{C_a}{\Sigma C}\left(1 + \frac{R_a}{\rho}\right)}.$$

It is improbable that R_a will exceed ρ in a practical circuit, in which case the effective gain will not drop appreciably below $g_m R_a$ and will be much less affected by variation of the chamber capacity C.

Appendix G also works out the gain of the circuit to a signal on the grid of the top valve. The ratio of the two gains shows how the shot noise of the bottom valve compares with the shot noise of the top valve at the output point of the circuit. This comparison is effected in Appendix G and is $\Sigma C/C_a$. Hence this circuit is identical, in respect of signal to shot noise ratio, with one using two triodes in a conventional cascaded amplifier as discussed earlier in this section.

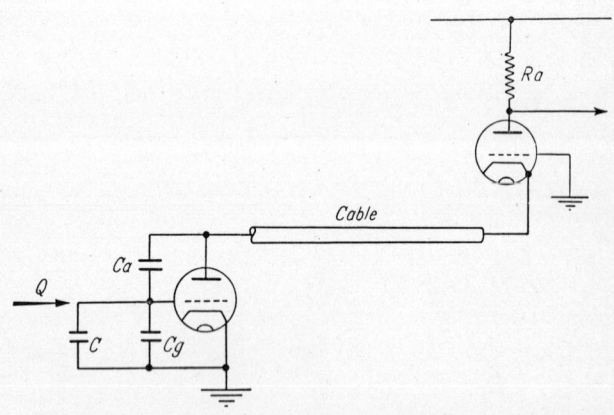

Fig. 28. Use of cascode with length of cable between valves

Since this circuit gives an overall gain which is usually of the order of $g_m R_a$ it is sometimes convenient to consider the whole as being roughly equivalent to a pentode in respect of gain but at the same time giving the signal to noise ratio of a triode. The circuit is known as the "cascode."

Another practical feature of the cascode which is sometimes very useful is shown in fig. 28.

The two valves can be separated by a reasonable length of cable, without affecting appreciably the performance of the circuit provided the capacitative reactance of the cable does not become comparable with $1/g_m$ of the top valve. This arrangement is advantageous where space considerations may allow of only a single valve being mounted direct on the ionisation chamber. The cascode is here superior in signal to shot noise ratio than any other circuit which could be used under the same conditions.

Before concluding this section on the relative merits of triodes and pentodes in the input stage of an ionisation chamber amplifier, it will now be shown that where the chamber capacity C is large compared with the valve capacity $C_g + C_a$ some improvement in the signal to shot noise ratio can be realised by connecting a number of valves in parallel. The analysis for the general case of n valves in parallel is given in Appendix H and the signal to shot noise ratio is $\dfrac{Q(n)^{1/2}}{v_{\text{tri}}(C + nC_g + nC_a)}$. This has a maximum value when $n = \dfrac{C}{C_g + C_a}$ showing that the optimum number of valves is reached when the total valve capacity is equal to the chamber capacity. Practically, however, the improvement obtained by paralleling valves is not very great relative to the effort involved, as the following figures show. Suppose the chamber capacity is three times the capacity $C_g + C_a$ of a single valve. If the signal to noise ratio of a single valve is represented by 1·0, the signal to noise ratio of two valves is 1·13 and the optimum signal to noise ratio of three valves is 1·16, — a total improvement of only 16%. The same improvement can be obtained by a reduction of the chamber capacity C by some 20% and this should obviously be the first line to investigate. The signal to flicker noise ratio is likewise improved by paralleling valves, but the signal to grid current noise ratio becomes considerably worse. This is clear, since the extra capacity that results from paralleling valves affects both the signal and the grid current noise of a single valve together, but there still remains the effect of the increased grid current itself to be considered. In the above example the signal to grid current noise ratio becomes $(3)^{1/2}$ times worse, due to the three to one increase in the grid current (r.m.s. noise $\propto (I_g)^{1/2}$). In general therefore, paralleling of input valves should only be used where grid current noise is insignificant and even then, only as a last resort.

4.3. RELATION BETWEEN SIGNAL TO NOISE RATIO, RESOLVING TIME AND AMPLIFIER BANDWIDTH

From equation (3) which gives the maximum amplitude of the signal reaching BB' in the equivalent circuit and from equations

(12), (15) and (18) which define the r.m.s. magnitudes of the main noise contributions at the same point, it is now possible to write down the following three equations for the signal to noise ratios. These are

$$\frac{S}{N}_{\text{(Shot)}} = \frac{V_0 T_1}{T} \frac{\left[\varepsilon^{T/T_1} - 1\right]^{\frac{T_1}{T_1 - T_2}}}{\left[\varepsilon^{T/T_2} - 1\right]^{\frac{T_2}{T_1 - T_2}}} \div \left[kT_e \frac{2 \cdot 5}{g_m} \frac{T_1}{T_2(T_1 + T_2)}\right]^{1/2} \quad (25)$$

$$\frac{S}{N}_{\text{(Grid current)}} = \frac{V_0 T_1}{T} \frac{\left[\varepsilon^{T/T_1} - 1\right]^{\frac{T_1}{T_1 - T_2}}}{\left[\varepsilon^{T/T_2} - 1\right]^{\frac{T_2}{T_1 - T_2}}} \div \left[\frac{eI_g}{2(\Sigma C)^2} \frac{T_1^2}{T_1 + T_2}\right]^{1/2} \quad (26)$$

and

$$\frac{S}{N}_{\text{(Flicker)}} = \frac{V_0 T_1}{T} \frac{\left[\varepsilon^{T/T_1} - 1\right]^{\frac{T_1}{T_1 - T_2}}}{\left[\varepsilon^{T/T_2} - 1\right]^{\frac{T_2}{T_1 - T_2}}} \div \left[\frac{10^{-13} T_1^2}{T_1^2 - T_2^2} \log_\varepsilon \frac{T_1}{T_2}\right]^{1/2}. \quad (27)$$

If initially in this section the analysis is restricted to cover a given chamber and input valve, i.e. T, g_m, and I_g can be regarded as constants, the signal to noise ratios vary with T_1 and T_2 as follows

$$\frac{S}{N}_{\text{(Shot)}} \propto \left[T_1 T_2\right]^{1/2} \left[T_1 + T_2\right]^{1/2} \frac{\left[\varepsilon^{T/T_1} - 1\right]^{\frac{T_1}{T_1 - T_2}}}{\left[\varepsilon^{T/T_2} - 1\right]^{\frac{T_2}{T_1 - T_2}}} \quad (28)$$

$$\frac{S}{N}_{\text{(Grid current)}} \propto \left[T_1 + T_2\right]^{1/2} \frac{\left[\varepsilon^{T/T_1} - 1\right]^{\frac{T_1}{T_1 - T_2}}}{\left[\varepsilon^{T/T_2} - 1\right]^{\frac{T_2}{T_1 - T_2}}} \quad (29)$$

and

$$\frac{S}{N}_{\text{(Flicker)}} \propto \left[\frac{T_1^2 - T_2^2}{\log_\varepsilon \frac{T_1}{T_2}}\right]^{1/2} \frac{\left[\varepsilon^{T/T_1} - 1\right]^{\frac{T_1}{T_1 - T_2}}}{\left[\varepsilon^{T/T_2} - 1\right]^{\frac{T_2}{T_1 - T_2}}}. \quad (30)$$

In addition the resolving time, given by equation (4) varies with T_1 and T_2 as follows

$$T_r \propto \frac{\left[\varepsilon^{T/T_2} - 1\right]^{\frac{T_2}{T_1-T_2}}}{\left[\varepsilon^{T/T_1} - 1\right]^{\frac{T_1}{T_1-T_2}}}. \qquad (31)$$

From an inspection of the right hand sides of expressions (28) to (31) inclusive, they are all unaltered if T_1 and T_2 are interchanged. Hence if T_1 and T_2 are allowed to vary in accordance with some relation which is not affected by interchange of T_1 and T_2 (e.g. $T_1 + T_2 =$ const. or $T_1 T_2 =$ const.) then any bandwidth having a ratio $T_1/T_2 = \theta$ will give rise to the same signal to shot noise, signal to grid current noise, and signal to flicker noise ratios, and the same resolving time as the corresponding bandwidth having a ratio $T_1/T_2 = 1/\theta$. Consequently, if the right hand sides of expressions (28) to (31) inclusive are computed and plotted for a range of values of T_1/T_2, subject to the first condition mentioned above, the resulting curves should be symmetrical about the point $T_1/T_2 = 1$. The law of variation relating T_1 and T_2 can be chosen at will. In this case it is convenient to take $T_1 T_2 =$ const. since this eases considerably the computation.

Suppose T_1 and T_2 are allowed to vary in accordance with

$$[T_1 T_2]^{1/2} = XT$$

where T is the chamber collection time and X is a constant relating the collection time T to the amplifier time constants T_1 and T_2. Suppose also $T_1/T_2 = \theta$ where θ is the independent variable. Expressions (28) to (31) inclusive can now be rewritten as

$$\frac{S}{N}_{(\text{Shot})} \propto \left[\theta\right]^{1/4} \left[1 + \frac{1}{\theta}\right]^{1/2} \frac{\left[\varepsilon^{\frac{1}{X}\frac{1}{[\theta]^{1/2}}} - 1\right]^{\frac{1}{1-\frac{1}{\theta}}}}{\left[\varepsilon^{\frac{1}{X}[\theta]^{1/2}} - 1\right]^{\frac{1}{\theta-1}}} \qquad (32)$$

$$\frac{S}{N}_{(\text{Grid current})} \propto \left[\theta\right]^{1/4} \left[1 + \frac{1}{\theta}\right]^{1/2} \frac{\left[\varepsilon^{\frac{1}{X}\frac{1}{[\theta]^{1/2}}} - 1\right]^{\frac{1}{1-\frac{1}{\theta}}}}{\left[\varepsilon^{\frac{1}{X}[\theta]^{1/2}} - 1\right]^{\frac{1}{\theta-1}}} \qquad (33)$$

$$\frac{S}{N}_{\text{(Flicker)}} \propto \left[\frac{\theta\left(1-\frac{1}{\theta^2}\right)}{\log_\varepsilon \theta}\right]^{1/2} \frac{\left[\varepsilon^{\frac{1}{X}\frac{1}{[\theta]^{1/2}}}-1\right]^{1-\frac{1}{\theta}}}{\left[\varepsilon^{\frac{1}{X}[\theta]^{1/2}}-1\right]^{\frac{1}{\theta-1}}} \qquad (34)$$

and
$$T_r \propto \frac{\left[\varepsilon^{\frac{1}{X}[\theta]^{1/2}}-1\right]^{\frac{1}{\theta-1}}}{\left[\varepsilon^{\frac{1}{X}\frac{1}{[\theta]^{1/2}}}-1\right]^{1-\frac{1}{\theta}}}. \qquad (35)$$

It will be noted that expressions (32) and (33) are now identical.

The right hand sides of expressions (32) to (35) inclusive have been computed for a number of specific values of X, in each case for a range of the independent variable θ from 0·1 to 10. It is found that the computed points for each expression do not vary for all values of X lying between 0·5 and ∞. Curves showing the percentage change in the signal to shot noise and signal to grid current noise ratios, in the signal to flicker noise ratio, and in the resolving time, valid for values of X between 0·5 and ∞, are shown in figs. 29 and 30. From these curves it can be seen that all the signal to noise ratios reach their maxima when $T_1 = T_2$, while the resolving time drops to a minimum. Hence it can be concluded that for values of X anywhere between 0·5 and ∞ the amplifier time constants T_1 and T_2 should be made equal, since this results in an improvement in the overall signal to noise ratio as well as in the resolving power of the system. Note that when $T_1 = T_2$ the ratio between the amplifier time constants and the chamber collection time is simply given by $T_1/T = X$.

Further computation shows that the general improvement in the signal to noise ratios and the resolving power is maintained, but to a lesser degree, for values of X down to 0·1. When X is very much less than 0·1 the signal to noise ratios deteriorate while the resolving time remains practically stationary. Hence it is not possible to conclude that T_1 should equal T_2 for X values less than 0·1, but since a chamber and amplifier working under such conditions would give a very poor signal to noise ratio anyhow on account of the serious attenuation of the signal by the amplifier time constants (see curves in fig. 9) this is a case of little practical interest and it is proposed to disregard it in the present discussion.

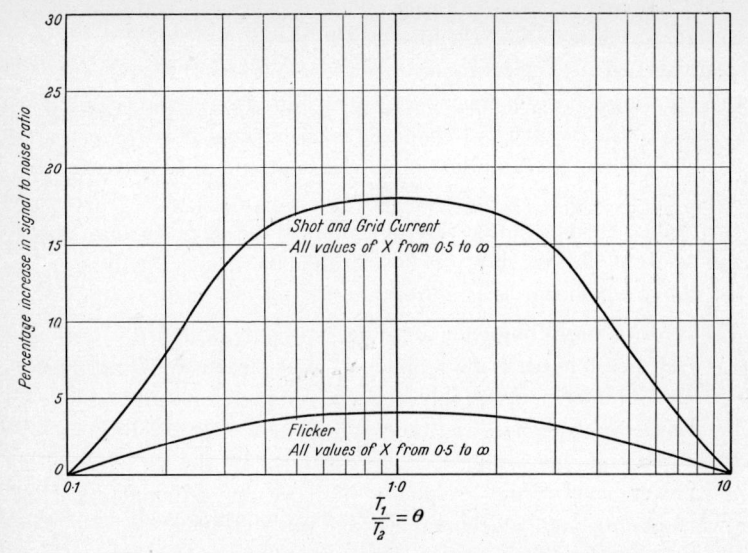

Fig. 29. Signal to noise ratio as a function of amplifier time constant ratio

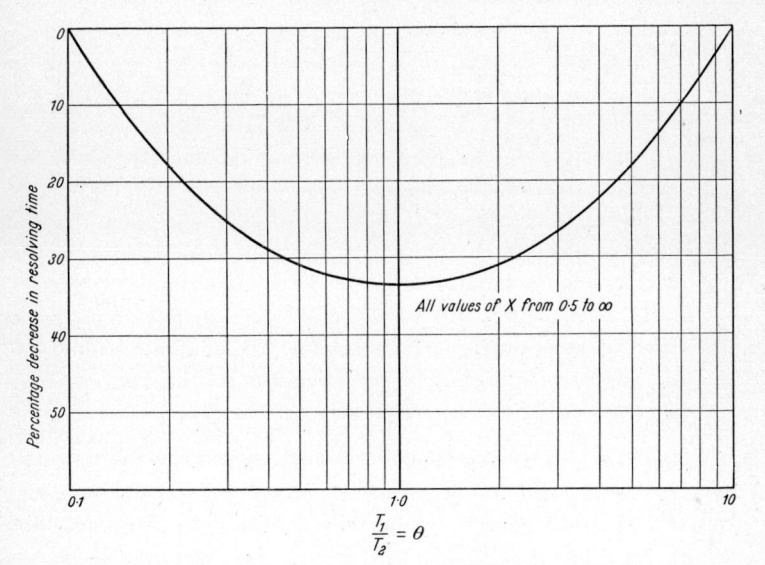

Fig. 30. Resolving time as a function of amplifier time constant ratio

Since I_g and g_m, although regarded as constants, are not limited as to their absolute values, the above analysis is applicable to any input valve and to any operating conditions, and it then becomes possible to state the general conclusion that in an ionisation chamber amplifier, to obtain the best signal to noise ratio consistent with a desired resolving time, the amplifier time constants of differentiation and integration should be made equal.

Although the percentage improvements in the signal to noise ratios, and resolving power, shown in figs. 29 and 30 are not very impressive, the above conclusion is an important one for two main reasons.

(a) In a high frequency amplifier, specially designed to give the best possible resolving power, the upper frequency limit of such an amplifier is invariably dictated by the slopes of the valves, the stray capacities and the required gain. An amplifier having equal time constants does not demand, for the same resolving power, such a high frequency limit as one with unequal time constants, and at the same time gives an improved signal to noise ratio. This is a point of some importance when extra bandwidth is difficult to achieve.

(b) So far the resolving time of the chamber and amplifier combination has been taken as the mean width of the pulse at the amplifier output. This simplified definition was used in Section 2.3, and also in the present section, on account of the difficulty in specifying an exact definition which would take into account all the relevant factors, namely, the chamber collection time T, the amplifier time constants T_1 and T_2, and the discriminator level. If, however, T_1 is made equal to T_2 it becomes possible to compute a series of curves showing the exact resolving time in terms of T, T_1 and the discriminator level. Such curves are very useful, since in any experiment which requires a certain resolving time, the amplifier bandwidth can be fixed immediately by reference to the curves. The analysis leading to the computation of the curves is as follows.

Equation (2) derived in Appendix A is a general expression for the pulse reaching BB' in the equivalent circuit, and valid for time T to ∞. It is also shown in Appendix A that the pulse maximum occurs at a time $t_m = \dfrac{T_1 T_2}{T_1 - T_2} \log_\varepsilon \dfrac{\varepsilon^{T/T_2} - 1}{\varepsilon^{T/T_1} - 1}$ and that this time is never less than the collection time T. Thus equation (2) taken over the time interval $\dfrac{T_1 T_2}{T_1 - T_2} \log_\varepsilon \dfrac{\varepsilon^{T/T_2} - 1}{\varepsilon^{T/T_1} - 1}$ to ∞ represents a general expression for the trailing edge of the pulse as it falls from V_m to zero. When $T_1 = T_2$ it is shown in Appendix I that equation (2) reduces to

$$V_{(T-\infty)} = \frac{V_0 T_1}{T} \varepsilon^{-t/T_1} \left[\varepsilon^{T/T_1} - 1 - \frac{T}{T_1} \varepsilon^{T/T_1} + \frac{t}{T_1}(\varepsilon^{T/T_1} - 1) \right] \tag{36}$$

while the time at which the maximum occurs is given by

$$t_m = \frac{T \varepsilon^{T/T_1}}{\varepsilon^{T/T_1} - 1}. \tag{37}$$

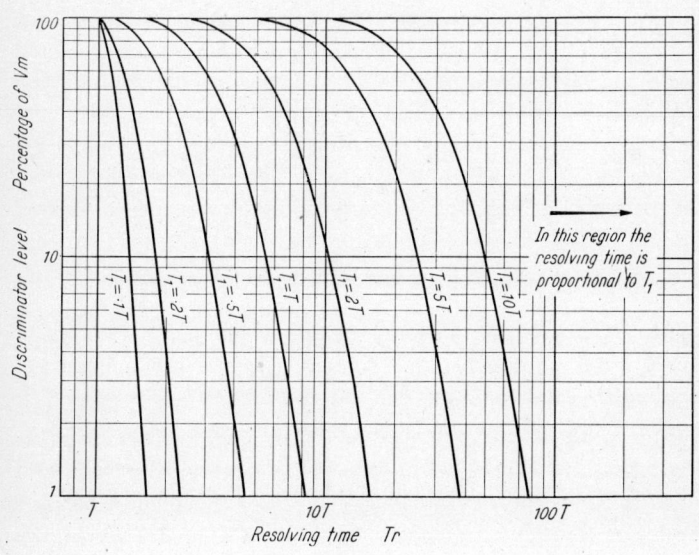

Fig. 31. Resolving time as a function of pulse amplitude and equal amplifier time constants

Equation (36) can now be used to compute a series of curves, each curve drawn for a particular value of T_1/T and each curve showing the time taken for the pulse to fall to specified percentages of V_m. This time is clearly the resolving time of the ionisation chamber and amplifier combination while the percentage of V_m defines the discriminator threshold level. The curves are shown in fig. 31 and apply for T_1/T values from 0·1 to 10. For values of T_1/T greater than 10 the resolving time is proportional to T_1, and can be obtained from the same curves by simple multiplication.

The use of the curves is probably best illustrated by an example. Suppose the chamber collection time T is 1 microsec., and that for a discriminator setting at 10% of the maximum output pulse the resolving time is required to be 10 microsec., i.e. $10T$. The curves show that T_1/T should be made equal to 2 and of course from previous conclusions $T_1/T_2 = 1$. This gives an amplifier frequency band defined by equal time constants of differentiation and integration, each equal to 2 microsec., and corresponding to a peaked frequency response curve centred on approximately 80 kc/s.

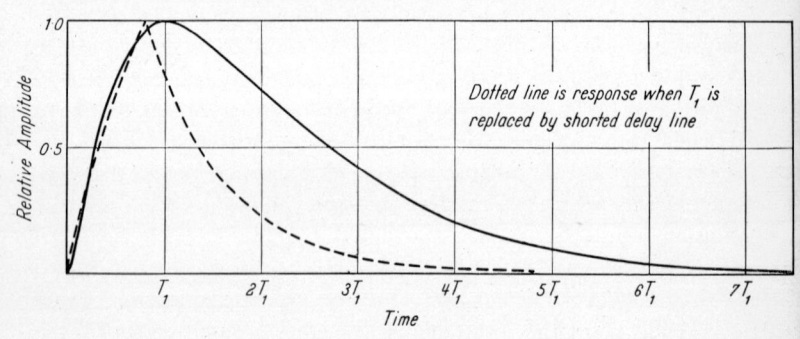

Fig. 32. Response of amplifier with $T_1 = T_2$ to a step input

When T_1/T is large, the shape of the output pulse is determined entirely by the amplifier time constants and is not influenced by the chamber collection time T. It is in fact the response of the amplifier to a step input and since this response is important and is used in subsequent sections of the text it will now be derived.

It is shown in Appendix J that when T_1/T is large equation (36) above reduces to

$$V_{(T-\infty)} = \frac{V_0 t}{T_1} \varepsilon^{-t/T_1} \tag{38}$$

and by differentiating this with respect to t and equating to zero, the time at which the maximum occurs is given by

$$t_m = T_1. \tag{39}$$

In the limit when T is zero, corresponding to a step input, equation (38) defines the response of the amplifier and this is shown in fig. 32 plotted to a time scale of T_1.

The absolute height of the pulse is from equation (38) equal to V_0/ε but for convenience in fig. 32 this has been represented as unity and the ordinate scale shows only relative amplitude.

4.4. CHOICE OF THE INPUT VALVE AND OPERATING CONDITIONS

Having fixed the amplifier frequency band to give the desired resolving time, the remaining problem is to choose the input valve and its operating conditions to give the best signal to noise ratio. For any valve and any bandwidth, the signal to noise ratio is determined by the relative magnitudes of the shot, grid current and flicker noise contributions. In general, the flicker noise contribution of most valves remains practically constant over a wide range of operating conditions, but the shot and grid current contributions, depending respectively on g_m and I_g, can vary appreciably with alteration of the anode current and anode voltage of the valve. In particular the shot noise can be improved at the expense of the grid current noise and vice versa. If the shot noise is initially greater than the grid current noise, then any alteration in the valve operating conditions which causes the mutual conductance to increase, will cause the total noise to decrease, this improvement continuing until the shot and grid current contributions are approximately equal. Beyond this point the total noise will increase again due to the grid current noise predominating. In general, therefore, the valve operating conditions should be adjusted to make the shot and grid current noise contributions equal, but only if this state of equality can be approached through a reduction in the larger component. This may not always be possible. For example, in an amplifier with a frequency band set to give a very short resolving time, a high slope valve should be used to keep down the shot noise. Even so, it may be found that with the highest slope valve available, the shot noise still predominates and this is a case where a state of equality can not be reached through a reduction in the larger component.

Amplifiers with high resolving power demand valves having a large mutual conductance. Typical examples are the EC91 and the 6AK5. Amplifiers with low resolving power demand valves with very small grid current. Although the EC91 and the 6AK5 can be run with reduced anode current and anode voltage, thereby giving much smaller grid current, it is found in practice that even better results can be obtained by changing to an alternative type of valve, usually one having a lower cathode dissipation. A typical example here is the ME 1400. It is also found in practice, and the reason for this will be made clear in the next section, that it is usually possible to obtain a better signal to noise ratio in an amplifier with low resolving power than in an amplifier with high resolving power—that is, signal to noise ratio and resolving power are opposing factors.

4.5. BEST POSSIBLE SIGNAL TO NOISE RATIO

In the previous section, the best signal to noise ratio consistent with a desired resolving time was considered. In this section it is proposed to discuss the best possible signal to noise ratio which can be achieved when there is no restriction whatsoever on the amplifier resolving time.

Consider a high slope valve running under such conditions that its mutual conductance is at its maximum limit. Suppose the amplifier frequency band is positioned in the spectrum to make the shot and grid current noise contributions equal. If the frequency band is now raised to a higher position in the spectrum, corresponding to an increased resolving power, the shot noise will also increase and thus the signal to noise ratio will get worse. Since the valve is already running at its maximum slope, it is not possible in this case to make any improvement in the signal to noise ratio by altering the operating conditions of the valve.

If, on the other hand, the frequency band is positioned lower in the spectrum, the grid current noise will increase, while the shot noise will decrease, but if at the same time, the valve operating conditions are altered to run with reduced anode current and/or anode voltage, the grid current noise will decrease again and it should be possible to bring the two components back to a state of equality. It is shown in Appendix K that if the I_g/g_m ratio under the new conditions is less than the I_g/g_m ratio under the initial conditions, the above process results in a reduction of the total noise and an improvement in the signal to noise ratio. In particular, if the ratio of the initial I_g/g_m to the new I_g/g_m is ϕ, then the improvement in the overall signal to noise ratio is $[\phi]^{1/4}$. This assumes that the flicker noise contribution is negligible, which, as will be apparent from later experimental results is nearly always the case. The smallness of the I_g/g_m factor of a valve is thus a measure of the best possible signal to noise ratio which can be realised, provided the frequency band is set to make the shot and grid current noise components equal.

Experiments show that for any given valve, the I_g/g_m factor does decrease as the anode current and anode voltage are lowered. For example, the ME 1400 with $V_a = 90$ volts and $I_a = 1$ mA has an $I_g/g_m = 1 \cdot 5 \times 10^{-7}$ while with $V_a = 45$ volts and $I_a = 150$ μA, $I_g/g_m = 2 \times 10^{-8}$. Consequently the signal to noise ratio can be improved by underrunning the valve but only at the expense of resolving power. This is the basis of the statement made in the previous section that signal to noise ratio and resolving power are opposing factors.

The decrease in the I_g/g_m factor does not continue indefinitely, however, as the grid current tends to assume a constant value due to the grid emission of electrons and the cathode emission of positive ions, while the mutual conductance keeps falling, so in general there is a minimum value for the I_g/g_m factor of a valve. It is to be expected, and this is also found in practice that valves with low cathode dissipation such as the ME 1400 exhibit lower I_g/g_m minima than valves dissipating a higher power. In every case however the minimum value for I_g/g_m invariably occurs when the mutual conductance and the grid current are extremely small and as a consequence the frequency band required to make the shot and grid current components equal is positioned very low in the spectrum. It can be concluded therefore, that the best possible signal to noise ratio is obtained by choosing and operating the input valve with the lowest possible value of I_g/g_m, and positioning the frequency band in the spectrum to make the shot and grid current noise contributions equal. This always results in an amplifier having very poor resolving power.

4.6. MINIMISING EFFECTS DUE TO VARIATION IN COLLECTION TIME

In many experiments for the determination of energy spectra, the location of the ionisation inside the chamber may vary slightly due to a spreading of the tracks of the energised particles. This gives rise to small variations in the collection time, which in turn may produce small changes in pulse amplitude at the amplifier output, depending on the relative magnitudes of the chamber collection time and the amplifier time constants. The pulse amplitude variations, if present, result in an extra widening of the spectrum peaks and a worsening of the signal to noise ratio.

The curves in fig. 9, if regarded in the light of T being the variable, show that when $T_1 = T_2$ the pulse amplitude is practically constant for all values of T from zero up to about T_1, and in this range any small variations in T will not produce corresponding variations in pulse amplitude. When T is greater than T_1, however, the pulse amplitude is dependent on T and in this range variations in T will give rise to corresponding variations in pulse amplitude. To minimise as far as possible any errors in pulse amplitude which may arise due to collection time variation, it is necessary to position the

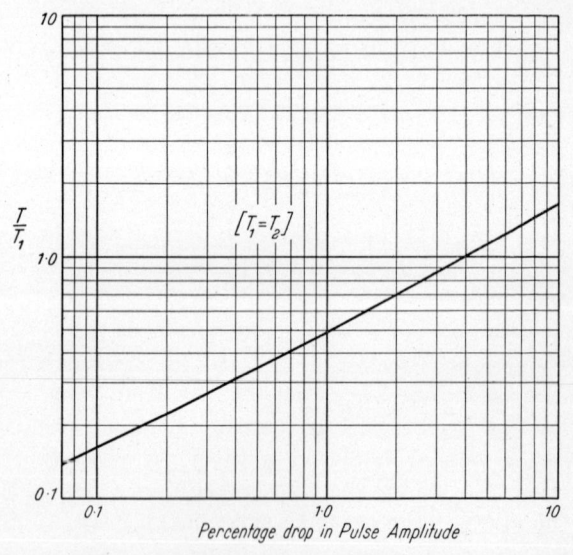

Fig. 33. Effect of collection time variation on pulse amplitude

frequency band sufficiently low in the spectrum to make the longest collection time preferably shorter than the amplifier time constants.

Since the curves in fig. 9 are not sufficiently accurate where small percentage changes are involved, a precise curve is drawn in fig. 33 for the case $T_1 = T_2$, and shows the percentage variation in pulse amplitude as the collection time T is increased with respect to the amplifier time constants.

For example, if it is required that the maximum possible variation in pulse amplitude should not exceed 0·5%* the curve indicates that

---

* In estimating the effect of this variation on the width of an energy peak the r.m.s. value is of interest. This depends on the geometry of the source and chamber and can not be easily defined. In the majority of cases however, one is probably safe in assuming the r.m.s. value to be at least three or four times less than the maximum variation.

T/T_1 should preferably be made less than 0·35. This corresponds to both the amplifier time constants at least three times the longest collection time of the chamber.

4.7. PULSE SHAPING USING A CRITICALLY DAMPED RINGING COIL

A relatively simple circuit element for pulse shaping, and one which is used extensively in the radar and television fields, is the ringing

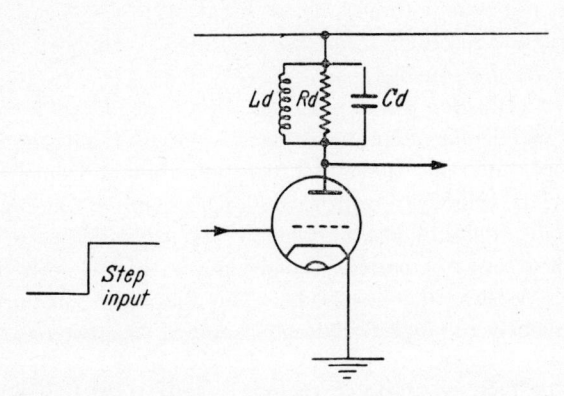

Fig. 34. Pulse shaping with critically damped ringing coil

coil with externally applied damping. In general circuit work the degree of damping may vary from critical damping on the first voltage swing to no damping on this swing and critical damping applied to the overswing through a thermionic diode or germanium rectifier. The damping required is usually adjusted to suit the particular circuit conditions. In pulse amplifier applications however, one of the chief criteria is that the pulse, when shaped, should have negligible overswing, so here critical damping on the initial swing must be employed.

In this section the ringing coil method of pulse shaping in a nuclear counter amplifier, is compared with that using equal time constants of differentiation and integration, particularly in respect of signal to noise ratio and resolving time. The circuit for analysis is shown in fig. 34.

For a step input of voltage applied to the grid of the valve, the transient analysis of the anode circuit shows that for critical damping the resistor Rd should be made equal to $\frac{1}{2}\left[\dfrac{L_d}{C_d}\right]^{1/2}$ and that subject to this condition the pulse on the anode varies with time according to $t\varepsilon^{\frac{-t}{2C_dR_d}}$. This is derived in Appendix L. It is immediately apparent that the pulse shape due to the critically damped ringing coil is similar to that produced by the equal time constant circuit and is in fact identical if $2C_dR_d$ is made equal to the differentiating time constant T_1. It follows that the pass band of the ringing coil amplifier must be the same as that of the equal time constant amplifier and thus both circuits are identical in respect of signal to noise ratio and resolving time, provided the two equality conditions stated above are satisfied.

The critically damped ringing coil circuit would thus appear to offer no particular advantages over a circuit using equal time constants of differentiation and integration. In fact, from a practical point of view, where bandwidth switching is required, the ringing coil circuit is at a definite disadvantage in that it is much more difficult to switch a series of parallel tuned circuits than to switch simple condenser resistor time constants. The dual time constant circuit is consequently to be preferred on account of its greater versatility.

4.8. PULSE SHAPING USING A SHORTED DELAY LINE

The use of a shorted delay line in general circuit work as a method of reducing the duration of a long pulse is well known. Since this is the function fulfilled by the differentiating time constant in an ionisation chamber amplifier, it is natural to enquire whether the shorted delay line offers any advantages over the simple condenser resistor differentiating circuit. Since a shorted delay line is a complex circuit element, comprising generally a series of coils and condensers, any advantages which it has over the simple differentiating circuit should be substantial in order to justify its use and the complexity involved.

The analysis in this section effects a comparison between a shorted delay line and a simple differentiating circuit in respect of signal to noise ratio and resolving time. To avoid undue complexity in the analysis the following assumptions are made:

(a) That the signal charge is collected instantaneously on the input capacity ΣC and consequently the amplifier receives a step input voltage on the first grid. This assumption eases the mathematics of the problem. The effect of a finite collection time, comparable with the amplifier time constants, is discussed briefly at the end of the analysis for zero collection time.

(b) That grid current and flicker noise can be neglected and that only shot noise need be considered. This is a permissible assumption, since for practical reasons, delay lines are not usually made with delay times in excess of 1 or 2 microsec. and thus the present application of a shorted line for pulse shaping is limited to high frequency amplifiers.

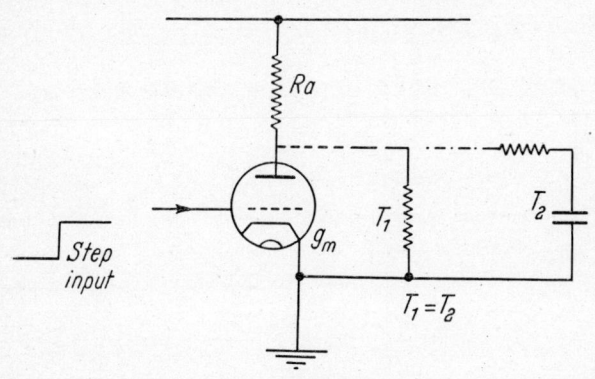

Fig. 35. Pulse shaping with equal time constants of differentiation and integration

To effect the comparison consider first the partial amplifier circuit shown in fig. 35, where the bandwidth is determined by equal time constants of differentiation and integration.

For a step input applied to the grid the pulse on the anode is proportional to $g_m R_a$. After passing through the time constants of differentiation and integration the maximum output pulse is then proportional to $\dfrac{g_m R_a}{\varepsilon}$.

The spectral density $\dfrac{v_s^2}{\delta f}$ of the shot noise on the anode is proportional to $[g_m R_a]^2$ and at the output is given by

$$\frac{v_s^2}{\delta f} \propto [g_m R_a]^2 \frac{\omega^2 T_1^2}{1 + \omega^2 T_1^2} \frac{1}{1 + \omega^2 T_2^2}$$

Therefore the mean square noise is

$$v_s^2 \propto [g_m R_a]^2 \int_0^\infty \frac{\omega^2 T_1^2 \, d\omega}{2\pi(1 + \omega^2 T_1^2)(1 + \omega^2 T_2^2)}$$

This integral has already been evaluated in Appendix C and is $\dfrac{1}{8T_1}$ or $\dfrac{1}{8T_2}$ when $T_1 = T_2$.

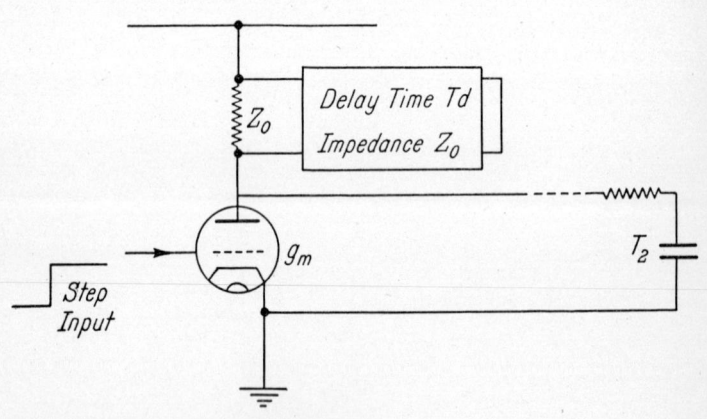

Fig. 36. Pulse shaping with shorted delay line and single time constant of integration

The signal to shot noise ratio of this time constant amplifier is therefore given by

$$\frac{S}{N}_{(\text{shot})} \propto \frac{\dfrac{1}{\varepsilon}}{\left[\dfrac{1}{8T_2}\right]^{1/2}} \qquad (40)$$

Consider now the circuit shown in fig. 36, in which the differentiating time constant T_1 has been replaced by a shorted delay line. The delay of the line from one end to the other is T_d and the characteristic impedance is Z_0. The end of the line connected to the anode of the valve is shunted by a resistive load equal to Z_0 to absorb the signal reflected from the shorted end.

In this case a step input applied to the grid gives a rectangular pulse on the anode of amplitude proportional to $\frac{g_m Z_0}{2}$ and lasting for a time $2T_d$. The output pulse appearing across the condenser of the integrating circuit rises exponentially with a time constant T_2 and reaches its maximum amplitude at time $2T_d$. Hence the maximum output pulse is proportional to $\frac{g_m Z_0}{2}\left[1 - \varepsilon^{-\frac{2T_d}{T_2}}\right]$

The spectral density $\frac{v_s^2}{\delta f}$ of the shot noise on the anode is proportional to $[g_m Z_0 \sin \omega T_d]^2$ where $[Z_0 \sin \omega T_d]^2$ represents the square of the anode impedance for all values of f. This is derived in Appendix M. Thus the spectral density at the output is given by

$$\frac{v_s^2}{\delta f} \propto [g_m Z_0]^2 \frac{\sin^2 \omega T_d}{1 + \omega^2 T_2^2}$$

and the total mean square noise is

$$v_s^2 \propto [g_m Z_0]^2 \int_0^\infty \frac{\sin^2 \omega T_d \, d\omega}{2\pi(1 + \omega^2 T_2^2)}$$

This integral is evaluated in Appendix N and is $\frac{1}{8T_2}\left[1 - \varepsilon^{-\frac{2T_d}{T_2}}\right]$

The signal to shot noise ratio of the delay line amplifier is therefore given by

$$\frac{S}{N}_{(\text{shot})} \propto \frac{\frac{1}{2}\left[1 - \varepsilon^{-\frac{2T_d}{T_2}}\right]^{1/2}}{\left[\frac{1}{8T_2}\right]^{1/2}} \qquad (41)$$

If now both amplifiers are required to give the same signal to noise ratio then

$$\frac{1}{\varepsilon} = \frac{1}{2}\left[1 - \varepsilon^{-\frac{2T_d}{T_2}}\right]^{1/2}$$

and solving this for T_d gives

$$T_d = 0\cdot 4\, T_2 \qquad (42)$$

Thus, if the differentiating time constant T_1 in an amplifier having $T_1 = T_2$ is replaced by a shorted delay line with a delay from one end to the other equal to 0·4 T_2, there will be no alteration in the signal to noise ratio. As far as alteration in the resolving time is concerned, this can be determined by comparing the pulse shapes at the amplifier output. The pulse shape for the dual time constant amplifier has already been derived and is shown in fig. 32. For the delay line amplifier the output pulse rises exponentially with a time constant T_2 during the time 0 to 0·8 T_2 and thereafter falls exponentially to zero with the same time constant. This pulse shape is shown by the dotted curve, also in fig. 32 for comparison purposes. It is clear that the delay line amplifier is almost twice as good as the time constant amplifier in respect of resolving power and since this is a substantial gain, it is concluded that the use of a shorted delay line for pulse shaping is to be recommended.

If the delay time T_d and the integrating time constant T_2 of the delay line amplifier are doubled, the dotted curve in fig. 32 is extended to twice its present duration and the trailing edge then practically coincides with the trailing edge of the time constant pulse. This, in effect, makes the resolving power of the delay line amplifier equal to the resolving power of the time constant amplifier. Equation (41) shows that this also results in a $[2]^{1/2}$ improvement in the signal to noise ratio of the former. Hence a delay line amplifier, having the same resolving power as a time constant amplifier, gives at the same time a 40% improvement in signal to noise ratio and demands an upper frequency limit only half as good as that required in the time constant amplifier. Put in this way, the advantages of delay line pulse shaping become even more significant.

In general, in ionisation chamber amplifiers, the conflicting requirements of signal to noise ratio and resolving power, only become serious in very high frequency amplifiers and it is here that pulse shaping using a shorted delay line has its chief application. Thus the assumption made at the beginning of this section to consider only shot noise and to neglect the grid current and flicker noise contributions is seen to be entirely validated by the results which have been obtained.

The foregoing remarks on delay line pulse shaping are only applicable to the case where the chamber collection time is small compared with the amplifier time constants. When they are of the same order of magnitude the problem becomes very involved in that the signal to noise ratio and the resolving time then depend on the collection time as well as on the bandwidth characteristics of the amplifier. No general solution would appear to be easily possible, but the author has investigated what can probably be regarded as the other extreme case, namely when the chamber collection time and the amplifier time constants are equal. It is not proposed to reproduce this analysis since much of it is graphical in nature but the results are of interest. It is found that the above conclusions in respect of signal to noise ratio and resolving time no longer hold and that there is in fact little or nothing to be gained

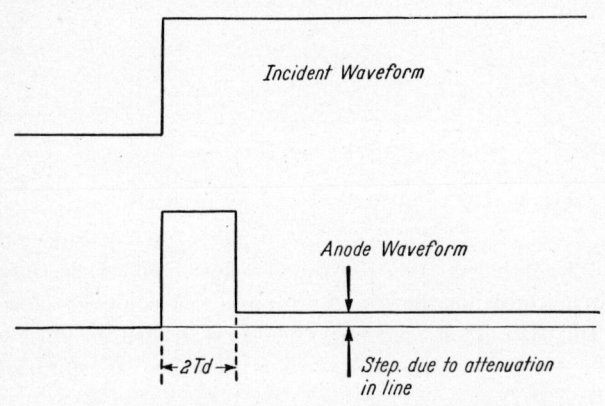

Fig. 37. Effect of attenuation in practical delay line

by adopting delay line pulse shaping in this case. The advantages of the delay line would seem therefore to depend on the factor T_1/T, being non-existent for this factor equal to unity and increasing to the limiting values derived above for T_1/T of the order of 10 or greater. This must always be borne in mind whenever delay line pulse shaping is contemplated.

In the practical application of a delay line to pulse shaping in an ionisation chamber amplifier, the following points must be observed—

(a) The cut off frequency of the line should be as high as possible and certainly considerably greater than the frequency corresponding to the integrating time constant T_2.

(b) The terminating resistor Z_0 should be matched as accurately as possible to the characteristic impedance of the line and care must be taken to ensure that there are no internal mismatches between sections of the line, otherwise the output pulse will have superimposed rings due to multiple reflections. (c) The resistive loss of the line should be kept as low as possible. The main effect of finite resistance in the line is to attenuate the wave front travelling along the line, with the result that the reflected wave does not completely cancel the incident wave and a step is produced in the anode waveform as shown in fig. 37.

This step can be removed by passing the anode waveform through a suitably chosen differentiating time constant as shown in fig. 38.

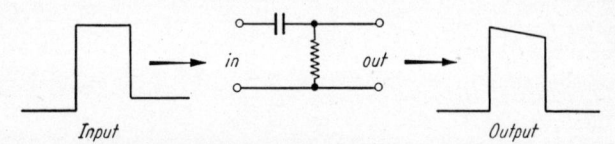

Fig. 38. Use of differentiating circuit to compensate for line attenuation

The differentiating time constant is chosen such that the droop which it produces on the top of the pulse just counterbalances the step at the base. If, for instance, the step is $\frac{1}{20}$th of the pulse amplitude, the time constant to give compensation is approximately $20\ (2T_d) = 40T_d$.

Because of the foregoing difficulties associated with a practical delay line, it is customary in general purpose amplifiers, where a wide selection of frequency bands is required, to use switched differentiating and integrating time constants, because of the extreme simplicity of this arrangement. A shorted delay line is generally only used in a high frequency amplifier, designed for a specific job, where the optimum of performance is required, and where one is prepared to take great care in the design, construction and operation of the line, in order to realise this performance.

5

SENSITIVITY

5.1. AMPLIFIER SENSITIVITY IN ENERGY MEASUREMENTS

In experiments for the determination of the energies of particles emitted from a radioactive source, high signal to noise ratios are the rule rather than the exception. The main problem is usually

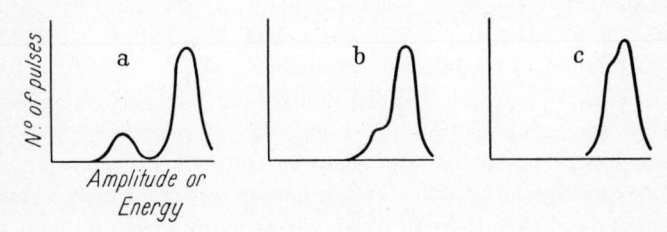

Fig. 39. Limitation in energy resolution caused by amplifier noise

that of distinguishing between pulses of nearly similar amplitude, corresponding to particles differing only slightly in their energies. The sensitivity of an ionisation chamber and pulse amplifier used for this type of work can therefore be measured by the minimum difference between two pulse amplitude groups which can be satisfactorily resolved by the kicksorter. Since the contour of an energy peak in the spectrum is determined usually by the r.m.s. noise of the amplifier, the sensitivity is primarily dependent on the magnitude of this noise factor. The sensitivity is also affected, to a lesser degree, by the relative heights of the two peaks which are to be separated. The height of any one peak corresponds to the activity of the substance emitting particles of that particular energy.

Fig. 39 shows typical kicksorter curves for two energy peaks, one having a height of about 4 times the other, as the separation between the peaks becomes smaller. The amplifier noise is assumed to be the same in all cases.

In fig. 39(a) the noise level of the amplifier is sufficiently low to show up both peaks clearly. In (b) the peaks are merging together but there is still no doubt that two peaks are present, while in (c) there is no more than a faint suspicion that a second energy peak is included.

With a curve such as (b) above the subsequent experimental procedure is to break up the composite curve into its two components. This is usually carried out graphically and is based on the assumption that the two peaks when separated will have similar Gaussian contours and that the r.m.s. value of each should be equal to that obtained with the standard calibrating source. Alternatively, the r.m.s. value can be obtained by measuring the amplifier noise level by the method described later in this section.* Such graphical separation may require a certain amount of trial and error before a good fit is obtained. With a curve such as (c) above the chance of this method succeeding will be small.

In order to try and establish a dividing line between possible success and probable failure using the graphical method, the author has proposed that the composite curve which gives a point of inflexion should be taken as the border between the two states. To one side of this, the two peaks will show up as in fig. 39(a) and the chances of separation will be high. To the other side the curve will be more like (c) and the chances of separation will consequently be low. This proposed dividing line also takes into account the relative height of the two peaks and its mathematical solution is given in Appendix O. The result is illustrated graphically in fig. 40.

This curve shows that if the two peaks have the same height, the amplifier pulses in each energy group should preferably differ by at least twice the amplifier r.m.s. noise level, if a reasonable chance of graphical separation is required. Again if the relative heights are in the ratio 10 to 1 then the pulses in the two energy groups should preferably differ by at least 3·35 times the r.m.s. noise.

---

* Any difference between the two r.m.s. values is an indication of additional fluctuations arising from the chamber or source. Fluctuations due to the chamber, or self absorption in the source can be reduced to negligible proportions by careful chamber design and source preparation. Fluctuations due to ionisation straggling are always present. These fluctuations are however less than normal and can usually be neglected for noise levels in excess of 500 ion pairs (see experimental results at end of this section).

SENSITIVITY 81

The prime factor determining the sensitivity of an ionisation chamber and pulse amplifier, used for energy analysis, is the magnitude of the r.m.s. noise level. It is customary to specify this magnitude relative to the input point of the amplifier—that is AA' in the equivalent circuit—and consequently the noise level is usually measured in terms of the signal charge at the input which produces a voltage signal at BB' in the equivalent circuit, equal to the total r.m.s. noise at the same point. It is convenient

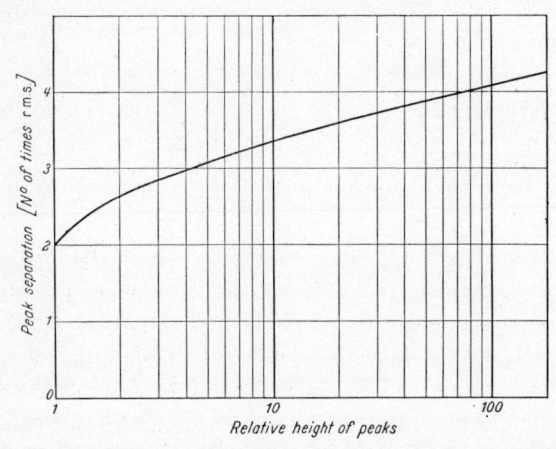

Fig. 40. Curve relating activities at two energy levels, the energy difference and the amplifier noise, to give a reasonable chance of separating the two energy groups

to assume that the signal charge is supplied instantaneously to the input capacity ΣC and then in any experiment where this is not the case—that is, the collection time T is comparable with or longer than the amplifier time constants—a correction can be made by referring to the curves in fig. 9. Since the voltage signal produced at the input depends on the total cold input capacity, this must also be stated along with the noise figures. As a final step of convenience, it is customary to measure the noise level not as an equivalent input charge of so many coulombs, but rather as a number of ion pairs where each ion pair corresponds to $1 \cdot 59 \times 10^{-19}$ coulombs. It is but one stage further to convert this to units of energy by multiplying by the ionising potential of the gas used in the chamber. Ionising potentials of the more commonly used gases range from about 24 to 36 eV.

If then v is the total r.m.s. noise voltage at BB' in the equivalent circuit, the voltage signal at the input for equality is $v\varepsilon$ allowing for the attenuation due to the amplifier time constants (see equation (38)). This corresponds to an input charge of $v\varepsilon\,\Sigma C$ coulombs and to $\dfrac{v\varepsilon\,\Sigma C}{1\cdot 59\times 10^{-19}}$ ion pairs. In all calculations of noise level or sensitivity this is the procedure adopted when using the noise expressions developed in Sections 3.2 to 3.5 inclusive.

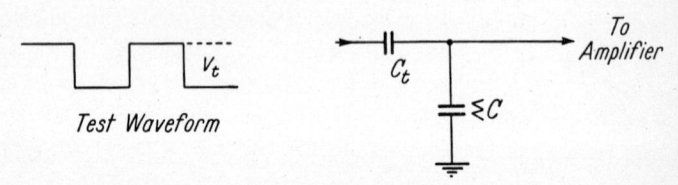

Fig. 41. Method of measuring charge sensitivity of ionisation chamber and pulse amplifier

To illustrate the foregoing remarks on sensitivity, consider the problem of separating the 238 and 235 isotopes of uranium. The former emits α particles of 4·20 MeV and the latter α particles of 4·35 MeV. The percentage abundance of the 238 isotope in naturally occurring uranium is 99·28% while that of the 235 isotope is 0·71%. The U238 half life on the other hand is $4\cdot 51\times 10^{9}$ years while the half life of the U235 is $8\cdot 52\times 10^{8}$ years. Thus the ratio of the activities is given by $\dfrac{99\cdot 28\times 8\cdot 52\times 10^{8}}{0\cdot 71\times 4\cdot 51\times 10^{9}}\doteqdot 26$. Fig. 40 shows that for a 26 to 1 ratio in the activities, the difference between the energy peaks should preferably be greater than 3·68 times the r.m.s. noise level of the amplifier. The energy difference is 150 KeV so the r.m.s. noise level of the amplifier should preferably be less $\dfrac{150}{3\cdot 68}=40\cdot 8$ KeV. Assuming a gas ionising potential of 30 eV this is equivalent to some 1360 ion pairs. This latter figure, together with the experimental noise level data for the proposed ionisation chamber and amplifier system, is sufficient information to decide whether the two uranium energy peaks can be easily separated or not.

To determine the chamber and amplifier noise level by experimental measurement, the following method is used.

A recurrent voltage waveform of variable but calibrated amplitude is fed through a very small condenser C_t to the amplifier input terminals as shown in fig. 41. The recurrence frequency of the voltage waveform should be such that its period is long compared with the pulse duration at the amplifier output and the voltage changes of the waveform should occur in a time preferably shorter than the amplifier time constants. The former requirement is to ensure that any pulse at the amplifier output, arising from an abrupt voltage change of the input waveform, is over before the arrival of a subsequent pulse. The latter requirement is to meet the condition that the charge should be supplied to the input capacity in a time short compared with the amplifier time constants.

As far as the test waveform is concerned the input capacity is C_t in series with ΣC and this is equal to $\dfrac{C_t \Sigma C}{C_t + \Sigma C}$. Hence during each abrupt voltage change of the input waveform, the charge alteration in both C_t and ΣC is $\dfrac{V_t C_t \Sigma C}{C_t + \Sigma C}$ and if C_t is very small compared with ΣC this reduces to $V_t C_t$. It is therefore possible to supply a known charge to the input capacity ΣC, no matter what value this may have, so long as the test voltage amplitude V_t and the test condenser C_t are known and so long as the latter is small compared with ΣC. The cold input capacity ΣC is never likely to drop below 10 $\mu\mu F$ in practice, hence if C_t is made a small fraction of $1\mu\mu F$ the above condition is satisfied and the charge is accurately known for all values of ΣC from $10\mu\mu F$ upwards. The control which alters the amplitude of the voltage waveform, can if desired, be calibrated directly in units of charge, or better still, ion pairs.

Using the above system the noise level of an ionisation chamber and pulse amplifier is obtained by measuring the total r.m.s. noise voltage at the amplifier output on a calibrated thermal instrument. Then the gain of the amplifier is reduced by a large known factor, of the order of 100, and the calibrated test signal fed to the input and adjusted until the pulse amplitude at the amplifier output is just equal to the previously measured r.m.s. noise level. This latter measurement can be made on an oscilloscope or by means of the amplitude discriminator which usually follows the amplifier. The r.m.s. noise level is then given by the test signal amplitude, expressed in ion pairs and divided of course by the attenuation factor which was introduced after the noise measurement.

The curve in fig. 42 shows the noise level of an EC91 input valve, measured as described above, and plotted as a function of the amplifier time constant T_1 ($T_1 = T_2$). The noise level is expressed in equivalent ion pairs across the total input capacity ΣC which

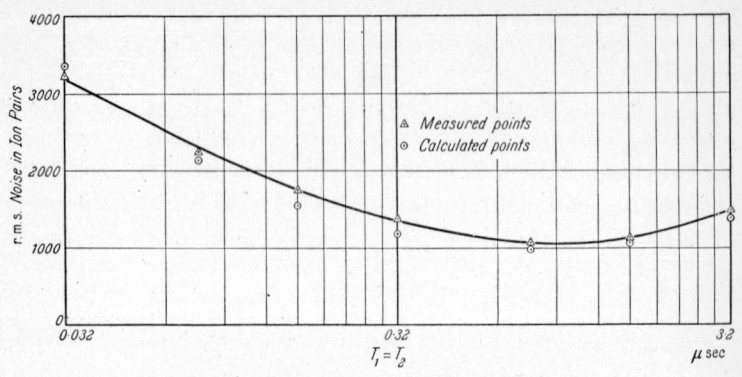

Fig. 42. Experimental noise level of EC91
$V_a = 250$ volts $\qquad g_m = 8\cdot 5$ mA/volt
$I_a = 10$ mA $\qquad\qquad \Sigma C = 46\ \mu\mu\mathrm{F}$
$I_g = 5 \times 10^{-8}$ amp
△ measured points ⊙ calculated points

in this case is $46\,\mu\mu F$. The input valve was run under the slope and grid current conditions mentioned in Section 3.4 and the corresponding noise level figures calculated from the valve parameters

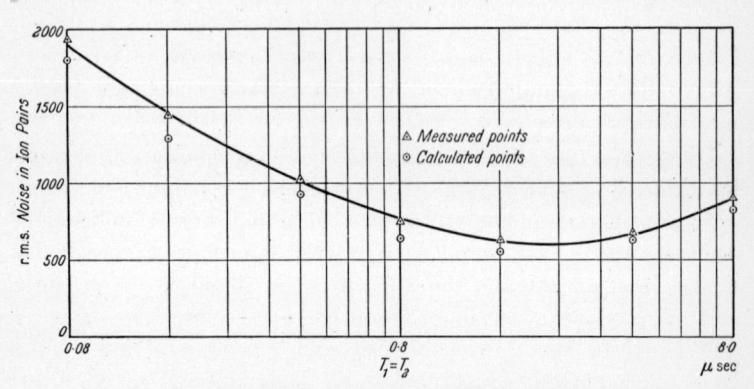

Fig. 43. Experimental noise level of 6AK5 (Triode connected)
$V_a = 120$ volts $\qquad g_m = 6\cdot 6$ mA/volt
$I_a = 10$ mA $\qquad\qquad \Sigma C = 34\ \mu\mu\mathrm{F}$
$I_g = 7 \times 10^{-9}$ amp
△ measured points ⊙ calculated points

are indicated by the ringed dots in fig. 42. It is seen that agreement between the calculated and experimental points is reasonably good.

SENSITIVITY

Fig. 43 shows the experimental curve for the 6AK5 valve, also run under the previously mentioned conditions. The noise level for low values of T_1 is slightly better than for the EC91, due primarily to the lower input capacity which in this case is $34\mu\mu F$. This reduced capacity more than offsets the lower slope of the 6AK5 valve. For larger values of T_1 the 6AK5 is considerably better than the EC91 and this is due entirely to the very much lower grid current of this valve. As before, agreement between the calculated and experimental points is reasonably good.

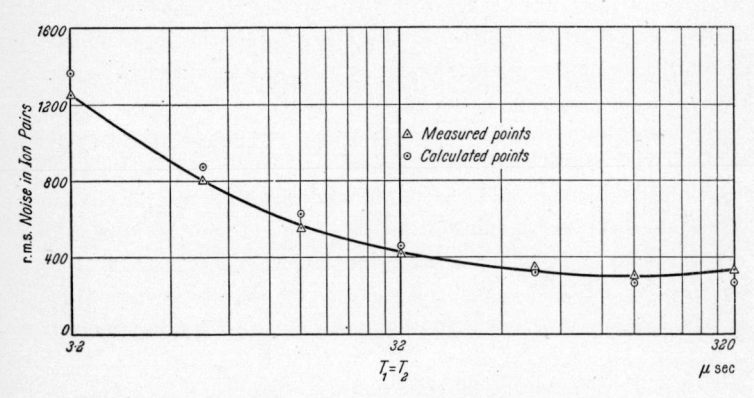

Fig. 44. Experimental noise level of ME 1400 (Triode connected)
$V_a = 45$ volts $\qquad g_m = 400\ \mu A/\text{volt}$
$I_a = 150\ \mu A \qquad \Sigma C = 40\ \mu\mu F$
$I_g = 8 \times 10^{-12}$ amp
△ measured points ⊙ calculated points

The experimental curve for the ME1400 with $V_a = 45$ volts and $I_a = 150\mu A$ is shown in fig. 44. Here the T_1 values are much larger than those for the previous two valves since the ME1400 is essentially intended for use in low frequency amplifiers. The total capacity in this case is $40\mu\mu F$. The measure of agreement between the calculated and experimental points is again reasonable, except perhaps at the longest time constant values where some difficulty was experienced in the measurement due to 50 c/s hum being picked up by the amplifier.

From an inspection of all three curves it is clear that there is a general reduction in the noise level as the amplifier time constants are lengthened, corresponding of course to a worsening of the resolving power of the amplifier. This again confirms the statement made in Section 4.4 that signal to noise ratio and resolving power are opposing factors.

The left hand side of each curve indicates a predominance of shot noise while the right hand side, especially for the EC91 and the 6AK5, indicates a predominance of grid current noise. Where shot noise is important, it is essential to state the total input capacity in all measurements of noise level, since the test signal required for noise equality varies as ΣC. In the case of the EC91 for instance, and for $T_1 = 0.032$ microsec., the r.m.s. noise in ion pairs would be 6400 for an input capacity of $92\mu\mu F$. Where grid current noise predominates on the other hand, the input capacity is relatively unimportant, unless it becomes sufficiently large to cause the shot noise to again exercise control.

The minimum point on each curve corresponds approximately to equality between the shot noise and the grid current noise components. One would therefore expect the minima of the three curves to bear some relation to the I_g/g_m factors of the valves. For the EC91, $I_g/g_m = 5.8 \times 10^{-6}$, for the 6AK5, $I_g/g_m = 1.06 \times 10^{-6}$ and for the ME1400, $I_g/g_m = 2 \times 10^{-8}$. By taking the fourth roots of the I_g/g_m ratios, one would expect the 6AK5 minimum to be about 1.5 times lower than that for the EC91 and the ME1400 minimum to be about 2.5 times lower than that for the 6AK5. This expectation is fairly well satisfied by the three minima of 1050, 600 and 300 ion pairs, especially when it is remembered that the input capacities are not the same in every case and that the flicker noise contribution is not negligible for the ME1400.

5.2. AMPLIFIER SENSITIVITY IN TOTAL ACTIVITY MEASUREMENTS

In counting experiments, where it is required to determine the total activity of an unknown source, high signal to noise ratios are the exception rather than the rule. As pointed out in Section 4.1, however, high signal to noise ratios are not essential and the majority of experiments can be satisfactorily carried out provided the signal to noise ratio, expressed in terms of the plateau length does not fall below about 2 or 3. In some experiments, however, where conditions prohibit signal to noise ratios even as high as 2, other factors must be taken into account, in particular the rate of counting of noise crests in relation to the rate of counting of genuine pulses. The sensitivity of an ionisation chamber and pulse amplifier used in such experiments can therefore be measured by the smallest signal which will allow of a known percentage of the signal pulses being counted, while, at the same time, the spurious counts due to noise crests do not exceed some specified maximum figure. As before, it is customary to measure the signal level in ion pairs relative to the amplifier input terminals and it is likewise assumed throughout this analysis that the signal charge is applied instantaneously to the input capacity ΣC. If this is not the case, the appropriate correction can be obtained from the curves in fig. 9.

The first step in the analysis is to examine the distribution of the noise amplitude with time, in order to derive a relation between the rate of counting of noise crests, the amplifier bandwidth and the discriminator trigger level. In so far as previous experimental results have shown that flicker noise is usually insignificant in comparison with the shot and grid current contributions, it is proposed to neglect it in this analysis and to consider only the latter two types of noise. Fundamentally, each type of noise arises from the random succession of a very large number of single events. In shot noise for instance each event is the transfer of an electron from the cathode to the anode of the valve, while in grid current noise, each event is the arrival of a positive ion at the grid of the valve. If $\bar{N}$ is the mean recurrence rate of events for a particular type of noise and $V(t)$ is the voltage effect at any point in the amplifier circuit arising from a single noise event, then the mean square noise voltage at that point is given by CAMPBELL's theorem[19, 20] and is

$$v^2 = \bar{N} \int_0^\infty \Big[V(t) \Big]^2 dt. \qquad (43)$$

Consider the case of grid current noise. If the grid current of the input valve is I_g then the number $\bar{N}$ of noise events per second is equal to I_g/e. Each event produces a step voltage across the input capacity ΣC of magnitude $e/\Sigma C$ and this in turn gives rise to a voltage $V(t)$ at BB' in the equivalent circuit of magnitude $\dfrac{e}{\Sigma C} \dfrac{t}{T_1} \varepsilon^{-t/T_1}$ when $T_1 = T_2$ (see equation (38)). The condition that $T_1 = T_2$ is of course representative of all ionisation chamber amplifiers where signal to noise ratio is important and it is hereafter assumed throughout this section that this condition is always satisfied. Insertion of the above values in equation (43) gives the mean square grid current noise voltage at BB' in the equivalent circuit as

$$v_g^2 = \frac{I_g}{e} \int_0^\infty \left[\frac{e}{\Sigma C} \frac{t}{T_1} \varepsilon^{-t/T_1} \right]^2 dt$$

$$= \frac{eI_g}{\Sigma C^2 T_1^2} \int_0^\infty t^2 \varepsilon^{-\frac{2t}{T_1}} dt.$$

The integral is evaluated in Appendix P and is $\dfrac{T_1^3}{4}$. Hence

$$v_g^2 = \frac{eI_g T_1}{4\Sigma C^2}$$

which is identical with equation (15) when $T_1 = T_2$.

The shot noise case is more difficult especially if space charge smoothing is considered[20]. In the present analysis, however, the space charge smoothing can be neglected, since this, although effecting a reduction in the r.m.s. noise level, does not affect the amplitude with time distribution of the noise since this is usually expressed in terms of the r.m.s. amplitude.

If the mean anode current of the valve is I_a then the number $\overline{N}$ of noise events per second is I_a/e Here, each noise event is the transit of an electron from the anode to the cathode of the valve and in an idealised case this can be considered as a pulse of anode current of magnitude e/τ where τ is the electron transit time. The voltage pulse on the grid of the valve, corresponding to the pulse of anode current is thus $e/g_m\tau$ and for convenience this can be regarded as the difference between two step voltages each of magnitude $e/g_m\tau$ but with one delayed a time τ with respect to the other. This is shown in fig. 45.

The voltage pulse $V(t)$ at BB' in the equivalent circuit therefore comprises two parts. One part, given by $\dfrac{e}{g_m T_1 \tau} t\varepsilon^{-t/T_1}$ is valid for

SENSITIVITY

time 0 to τ while the second part, given by

$$\frac{e}{g_m T_1 \tau}\left[t\varepsilon^{-t/T_1} - (t-\tau)\varepsilon^{-\frac{(t-\tau)}{T_1}}\right]$$

is valid from τ to ∞. If τ is small compared with T_1, and this applies for all frequencies under consideration in the text, the first part of $V(t)$ is approximately equivalent to a linear rise of slope $e/g_m T_1 \tau$

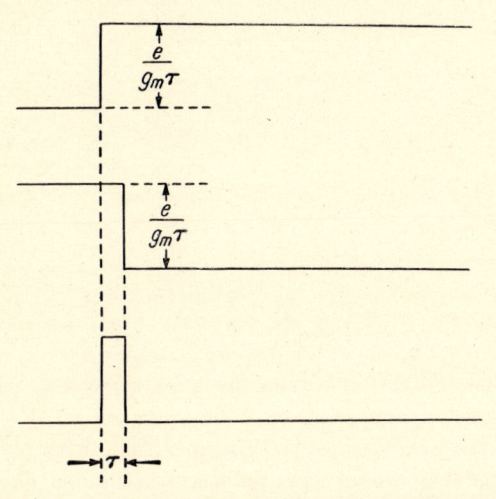

Fig. 45. Representation of a short pulse as the difference between two step functions

during the time 0 to τ while the second part, valid from τ to ∞ is approximately equal to $\dfrac{e}{g_m T_1}\dfrac{d}{dt}\left[(t-\tau)\varepsilon^{-\frac{(t-\tau)}{T_1}}\right]$

This gives

$$V(t) = \frac{et}{g_m T_1 \tau} + \frac{e}{g_m T_1}\varepsilon^{-\frac{(t-\tau)}{T_1}}\left(1 - \frac{t-\tau}{T_1}\right)$$

from 0 to τ \qquad from τ to ∞

Hence the mean square shot noise voltage at BB' in the equivalent circuit is given by

$$v_s^2 = \frac{I_a e}{g_m^2 T_1^2}\left[\int_0^\tau \frac{t^2}{\tau^2}dt + \int_\tau^\infty \varepsilon^{-\frac{2(t-\tau)}{T_1}}\left(1 - \frac{t-\tau}{T_1}\right)^2 dt\right]$$

Again, if τ is small compared with T_1, the first integral can be neglected in comparison with the second. Also the second integral is clearly the same as

$$\int_0^\infty \varepsilon^{-\frac{2t}{T_1}} \left(1 - \frac{t}{T_1}\right)^2 dt.$$

This is evaluated in Appendix Q and is $\dfrac{T_1}{4}$.

Thus
$$v_s^2 = \frac{I_a e}{4 g_m^2 T_1}$$

This expression is the same as that which would be obtained by integrating equation (9) between the frequency limits set by T_1 and T_2 and then introducing the conditions $T_1 = T_2$ and $F^2 = 1$.

To determine the rate of counting of noise crests by an amplitude discriminator set at a level x/v times the r.m.s. noise the analysis is as follows[21]. The discriminator will trigger every time the noise waveform passes through the threshold level x in the positive direction, thus what is required is the number of times per second the noise waveform reaches the trigger level x with positive slope.

The probability that the noise just reaches the trigger level x is $\dfrac{1}{(2\pi)^{1/2} v} \varepsilon^{-\frac{1}{2}(x/v)^2}$ where v is given by equation (43). This probability is infinitesimal and what is of more interest in the present application is the probability that at some time $t - dt$ say, the noise waveform lies within the limits x and $x - dx$. This is given by

$$p_{(x,\, x-dx)} = \frac{1}{(2\pi)^{1/2} v} \varepsilon^{-\frac{1}{2}\left(\frac{x}{v}\right)^2} dx \qquad (44)$$

Now the amplitude of the noise waveform at any instant is due to the direct addition of the amplitudes $V(t)$ of the individual noise events. Likewise it can be assumed that the velocity of the noise waveform at any instant is due to the direct addition of the velocities $\dot{V}(t)$ of the individual noise events, where $\dot{V}(t) = \dfrac{d}{dt} V(t)$. Thus, the probability that the noise waveform velocity at any instant should lie between $\dot{x}$ and $\dot{x} - d\dot{x}$, is $\dfrac{1}{(2\pi)^{1/2}\,\dot{v}}\,\varepsilon^{-\frac{1}{2}\left(\frac{\dot{x}}{\dot{v}}\right)^2} d\dot{x}$ where by analogy with equation (43)

$$\dot{v}^2 = \overline{N} \int_0^\infty [\dot{V}(t)]^2\, dt \tag{45}$$

It follows that the probability that the noise waveform should have a positive velocity at any instant is given by

$$p_{(\text{vel. +})} = \frac{1}{(2\pi)^{1/2}\,\dot{v}} \int_0^\infty \varepsilon^{-\frac{1}{2}\left(\frac{\dot{x}}{\dot{v}}\right)^2} d\dot{x} \tag{46}$$

Combining equations (44) and (46) gives the probability that the noise waveform at the time $t - dt$ should lie between x and $x - dx$ and at the same time should have a positive velocity. This is

$$p_{(x,\,x-dx)} \times p_{(\text{vel. +})} = \frac{1}{2\pi v \dot{v}}\,\varepsilon^{-\frac{1}{2}\left(\frac{x}{v}\right)^2} dx \int_0^\infty \varepsilon^{-\frac{1}{2}\left(\frac{\dot{x}}{\dot{v}}\right)^2} d\dot{x}.$$

Finally if it is specified that the noise waveform should reach x not later than time t, this condition can be satisfied if dx is made equal to $\dot{x}\,dt$. Thus the probability p_x that the noise waveform should pass through the level x with positive velocity is

$$p_x = \frac{dt}{2\pi v\dot{v}}\,\varepsilon^{-\frac{1}{2}\left(\frac{x}{v}\right)^2} \int_0^\infty \dot{x}\,\varepsilon^{-\frac{1}{2}\left(\frac{\dot{x}}{\dot{v}}\right)^2} d\dot{x}.$$

This is a definite integral which is tabled in most text books on calculus and is equal to $\dot{v}^2$.
Hence

$$p_x = \frac{\dot{v}\,dt}{2\pi v}\,\varepsilon^{-\frac{1}{2}\left(\frac{x}{v}\right)^2}$$

from which it follows that the number P of noise crests counted per second for a discriminator level x/v times the r.m.s. noise is

$$P = \frac{\dot{v}}{2\pi v}\,\varepsilon^{-\frac{1}{2}\left(\frac{x}{v}\right)^2}. \tag{47}$$

The factor $\dot{v}/2\pi v$ gives the number of times per second the noise waveform crosses the zero axis in the positive direction—that is, the number of noise counts per second for zero discriminator level—while the factor $\varepsilon^{-\frac{1}{2}\left(\frac{x}{v}\right)^2}$ shows how this varies with the setting of the discriminator relative to the r.m.s. noise. This latter factor is plotted in fig. 46 as a function of x/v.

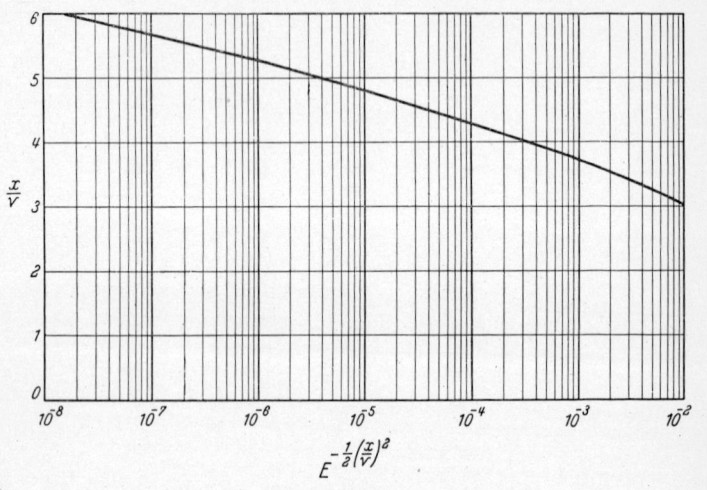

Fig. 46. Variation in rate of counting of noise crests with discriminator level

The above result can now be applied to the two types of noise under consideration in this section. For grid current noise v_g^2 is proportional to $\int_0^\infty t^2 \varepsilon^{-\frac{2t}{T_1}} dt$ and this is equal to $T_1^3/4$ from Appendix P. Also $\dot{v}_g^2$ is proportional to

$$\int_0^\infty \left[\frac{d}{dt}\left(t\varepsilon^{-\frac{t}{T_1}}\right)\right]^2 dt = \int_0^\infty \varepsilon^{-\frac{2t}{T_1}}\left(1 - \frac{t}{T_1}\right)^2 dt$$

and this is equal to $T_1/4$ from Appendix Q. Hence the rate of counting of grid current noise crests is given by

$$P_g = \frac{1}{2\pi T_1} \varepsilon^{-\frac{1}{2}\left(\frac{x}{v}\right)^2}. \qquad (48)$$

For example, if the amplifier time constants T_1 and T_2 are both equal to 3·2 microsec., if the noise is entirely grid current noise and if the discriminator is set to a level of 4·65 times the r.m.s. noise, then there will be a noise counting rate of 1 per second.

For shot noise, v_s^2 is proportional to

$$\int_\tau^\infty \varepsilon^{-\frac{2(t-\tau)}{T_1}} \left(1 - \frac{t-\tau}{T_1}\right)^2 dt$$

and this is equal to $T_1/4$ from Appendix Q. Also $\dot{v}_s^2$ is proportional to

$$\int_0^\tau \left[\frac{d}{dt}\left(\frac{t}{\tau}\right)\right]^2 dt + \int_\tau^\infty \left[\frac{d}{dt}\left\{\varepsilon^{-\frac{(t-\tau)}{T_1}}\left(1 - \frac{t-\tau}{T_1}\right)\right\}\right]^2 dt.*$$

This is evaluated in Appendix R and is $\dfrac{1}{\tau} + \dfrac{5}{4T_1}$. Hence the rate of counting of shot noise crests is given by

$$P_s = \frac{1}{2\pi}\left[\frac{\dfrac{1}{\tau} + \dfrac{5}{4T_1}}{\dfrac{T_1}{4}}\right]^{1/2} \varepsilon^{-\frac{1}{2}\left(\frac{x}{v}\right)^2}$$

$$= \frac{1}{2\pi T_1}\left[\frac{4T_1}{\tau} + 5\right]^{1/2} \varepsilon^{-\frac{1}{2}\left(\frac{x}{v}\right)^2}. \qquad (49)$$

Comparing equation (49) for shot noise with equation (48) for grid current noise, it will be seen that for a given discriminator level relative to the r.m.s. noise, the shot noise counting rate is greater than the grid current noise counting rate. This is not an unreasonable finding, but further inspection shows that the ratio between the two noise counting rates is not constant, as might be expected, but depends on the factor T_1/τ. This latter finding is difficult to accept, since experimental measurement of the counting rate of shot noise would tend to confirm that this varies inversely as T_1 in the same way as grid current noise. The disagreement is almost certainly due to differences between the theoretical amplifier

---

* Whereas in the calculation of v_s^2 it was possible to neglect the integral appertaining to that part of $V(t)$ between 0 and τ, this is no longer possible in the calculation of $\dot{v}_s^2$ since here the integral corresponds to a large, and as will be seen, important velocity term.

and discriminator considered above and the practical equipment used for the tests. In equation (49) for instance, the two components which add together to give the theoretical rate of counting of shot noise crests refer respectively to fluctuations associated with the leading and trailing edges of the primary shot noise impulse. Because of the very large difference between the time durations of these two edges, the noise waveform will consist in the main of long term fluctuations due to the trailing edge of the pulse together with numerous superimposed short term fluctuations of much smaller amplitude due to the leading edge of the pulse. A practical amplitude discriminator, receiving such a noise waveform will tend to disregard all, or a large proportion of the short term small amplitude fluctuations due to such effects as backlash and finite triggering time of the discriminator. On this basis it would appear that a more representative form of equation (49) might be

$$P_s = \frac{[5]^{1/2}}{2\pi T_1} \varepsilon^{-\frac{1}{2}\left(\frac{x}{v}\right)^2}. \qquad (50)$$

The problem is further complicated however by the fact that a practical amplifier is not an infinite bandwidth device containing only a single time constant of integration. Other integrating time constants are unavoidably present and although these may be very much smaller than the main integrating time constant they are nevertheless much longer than the rise time τ of the theoretical noise pulse. The effect of this is to slow down appreciably the leading edge of the primary noise pulse and also to convert this pulse into a continuous function of time, more akin to the grid current noise impulse, as distinct from the two part impulse considered in the foregoing theoretical analysis. There seems to be no simple and obvious way of allowing for this in equation (49) and it would appear that some alternative method of approach is called for. The alternative method adopted by the author is as follows. Each type of noise has associated with it a time interval Δ say, hereafter known as the response time, during which the r.m.s. fluctuation in the number of noise events occurring in this time Δ is equal to the r.m.s. amplitude fluctuation of the noise expressed in terms of the maximum amplitude of a single event. In grid current noise for instance $v_g^2 = \dfrac{eI_g T_1}{4\Sigma C^2}$ while the maximum amplitude of a single event is $\dfrac{e}{\Sigma C}\dfrac{1}{\varepsilon}$ and the number of noise events per second is of course I_g/e.

Thus

$$\left[\dfrac{I_g}{e}\Delta_g\right]^{1/2} = \left[\dfrac{eI_gT_1}{4\Sigma C^2}\right]^{1/2}\left[\dfrac{\Sigma C\varepsilon}{e}\right]$$

giving the response time for grid current noise as

$$\Delta_g = \dfrac{\varepsilon^2 T_1}{4}$$

$$= 1\cdot 85\, T_1. \tag{51}$$

With shot noise on the other hand $v_s^2 = \dfrac{I_a e}{4g_m^2 T_1}$ while the maximum amplitude of a single event is $e/g_m T_1$ and the number of noise events per second is I_a/e.

Thus

$$\left[\dfrac{I_a}{e}\Delta_s\right]^{1/2} = \left[\dfrac{I_a e}{4g_m^2 T_1}\right]^{1/2}\left[\dfrac{g_m T_1}{e}\right]$$

giving the response time for shot noise as

$$\Delta_s = \dfrac{T_1}{4}$$

$$= 0\cdot 25\, T_1. \tag{52}$$

For a given value of T_1 the response time of each type of noise is an approximate measure of the duration of the crests associated with that type of noise. Since the noise counting rate varies inversely with the duration of the noise crests it follows that

$$\dfrac{P_s}{P_g} = \dfrac{\Delta_g}{\Delta_s} = 7\cdot 4$$

and therefore this analysis gives the rate of counting of shot noise crests as

$$P_s = \dfrac{7\cdot 4}{2\pi T_1}\varepsilon^{-\frac{1}{2}\left(\frac{x}{v}\right)^2}. \tag{53}$$

In the above calculation of Δ_s, the maximum amplitude of a single noise pulse has been taken as $e/g_m T_1$. Although this refers to the amplitude of a noise pulse after passing through a theoretical amplifier it is also a reasonable measure of the noise pulse emerging from a practical amplifier since considerable slowing down of the leading edge by unavoidable integration has only a minor effect on the pulse amplitude, provided the extra integration does not become comparable with that due to the main integrating time constant. Thus, there would seem some reasonable justification for using equation (53) to give the rate of counting of shot noise crests in a practical case in preference to equations (49) or (50). This is further supported by experimental measurements on shot noise to be presented later in this section. It must be stated here and now however, that this experimental work is incomplete and that much remains to be done before all doubtful factors associated with the practical rate of counting of shot noise crests are completely removed. The author's work is only in the nature of a beginning. It is included, however, because as far as it goes it supports equation (53), it will give considerable guidance to anyone contemplating a more detailed investigation of this problem, and should further work prove equation (53) to be incorrect this will only affect some of the author's intermediate conclusions, the final results given in figs. 48 and 49 at the end of this section will still remain valid.

Assuming then that equation (53) holds and taking as before a numerical example, if the amplifier time constants T_1 and T_2 are both equal to 3·2 microsec, if the noise is entirely shot noise and if the discriminator is set to a level of 5·07 times the r.m.s. noise then there will be a noise counting rate of 1 per second. For the same bandwidth and noise counting rate, the corresponding x/v figure for grid current noise is 4·65. There is therefore a ratio of
$$\frac{5 \cdot 07}{4 \cdot 65} = 1 \cdot 09$$
between the two discriminator levels for shot and grid current noise respectively.

So far the analysis has been directed towards finding a relationship between the rate of counting of noise crests, the amplifier bandwidth and the discriminator level. Bearing in mind the sensitivity definition given at the beginning of the present section, it now remains to introduce into the analysis the amplitude of the signal required in order that a known proportion of the signal pulses can be counted under the stipulated noise and bandwidth conditions.

The full line curves in fig. 47 show how the proportion of signal pulses counted varies with the amplitude of the signal pulses. One curve refers to grid current noise only and the other to shot noise only.

The curves were obtained experimentally in the following way. The amplifier time constants T_1 and T_2 were each set to 3·2 microsec, the input valve used was an EC91 having a high order of grid current

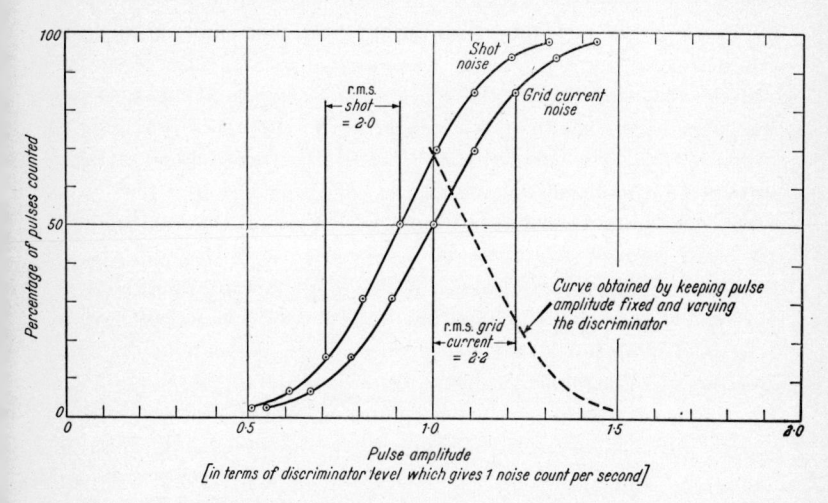

Fig. 47. Experimental relationship between percentage of pulses counted and pulse amplitude, for shot noise and grid current noise

to ensure that all the noise at the amplifier output was grid current noise and the amplitude discriminator was set to give a noise count of 1 per second. Test pulses of a high and known recurrence frequency (several hundred per second) were then fed into the amplifier by the method described in Section 5.1 and the proportion of these actuating the discriminator was measured on a ratemeter[22], this instrument being more convenient than a scaler and register because of its direct indicating facility. Readings were taken of pulse amplitude against the proportion of pulses counted and from a subsequent calibration check on the equipment, the pulse amplitude was expressed in terms of the discriminator level corresponding to 1 noise count per second. This permitted the grid current curve in fig. 47 to be drawn. The shot noise curve was obtained in the same way, except that the input valve was replaced by an ME1400 to ensure that the noise was predominantly shot noise. The discriminator was again set to the level giving 1 noise count per second and the results are shown by the shot noise curve in fig. 47. A number of extremely important conclusions can be obtained from these experimental curves.

In the first place the shape of each curve would appear to define accurately the amplitude distribution of the noise.* To be more precise the amplitude distribution of the noise should be obtained by keeping the test pulse amplitude fixed and moving the discriminator trigger level. This was checked for the shot noise case and the resulting curve is shown dotted in fig. 47. It will be seen that this is almost an exact image of the shot noise curve from which it may be concluded that the full line curves do show fairly accurately the noise distribution. The ringed dots adjacent to each curve have been plotted from tables giving the area under a Gaussian contour and the points chosen to give the best overall fit with the experimental curves. The good measure of agreement here shows how well the experimental curves do in fact follow the type of distribution which has been assumed throughout the text. It is possible therefore to derive from each curve the value of the r.m.s. noise in relation to the discriminator level and for grid current noise this is 0·22 giving a ratio $\frac{\text{Discriminator Level}}{\text{r.m.s. noise}} = 4\cdot55$. This should be compared with the theoretical value of 4·65. For shot noise on the other hand, the r.m.s. value is 0·2 giving a ratio $\frac{\text{Discriminator Level}}{\text{r.m.s. noise}} = 5\cdot0$ for comparison with the theoretical value of 5·07. Agreement in each case is within 3% and would appear to confirm the reasonable accuracy of equations (48) and (53) developed earlier in this section.

The second important observation from the curves in fig. 47 is that for grid current noise only, signal pulses of amplitude just equal to the discriminator level give a 50% count whereas for shot noise the corresponding percentage is 68%. This result is reasonable, when it is remembered that the grid current noise crests have the same general shape and duration as the signal pulses whereas the

---

* Here the distribution curve is the integral of the Gaussian curve shown in fig. 23. This is because the single amplitude discriminator used in the experiment counts all pulses which reach and also exceed the trigger level. The total area under the Gaussian contour is usually represented by unity, or in this case 100%.

shot noise crests are considerably shorter*. To reduce the percentage of signal pulses counted in the shot noise case from 68% to 50% the curve in fig. 47 shows that pulses of amplitude 0·91 times the discriminator level are required. When it is remembered that the shot noise discriminator level for 1 noise count per second is 1.09 times the corresponding level for grid current noise, these two effects approximately cancel each other and it follows that for both grid current noise and shot noise, the same signal amplitude relative to the r.m.s. noise is required to give a 50% count when the discriminator is set to give 1 noise count per second.

So far this fortuitous cancellation corresponds only to the ratio between the 5·07 and the 4·65 x/v points on the curve in fig. 46. If the amplifier bandwidth or the required noise counting rate are altered these two points will move up or down this curve, the abscissae ratio between them will remain constant and equal to $P_s/P_g = 7·4$, but the ratio of their ordinates will vary above and below the value 1·09. It is therefore necessary to investigate whether the cancellation effect, derived from fig. 47, keeps in step with the variation in the ratio of the two discriminator levels.

Experiments were carried out on the shot noise curve in fig. 47 at discriminator levels 1·5 and 2 times the value used in the original experiment. In the former case it was found that signal pulses of amplitude just equal to the discriminator level gave a 63% count, and in the latter case the percentage count was 59%. The two curves apparently get closer together for higher discriminator levels relative to the r.m.s. noise and although it was not possible to check this the converse is probably true. Again, this is a reasonable result since the top of the signal pulses get narrower relative to the duration of the shot noise crests as the signal to noise ratio increases and thus the signal pulses counted should tend to approach the 50% value.

From the above experimental figures, it would appear that for shot noise and grid current noise discriminator levels different from the values relating to the curves in fig. 47, that is for noise

---

* Throughout this and subsequent experiments a paralysis time slightly longer than the duration of a signal pulse was applied to the discriminator each time it was triggered. This was to make certain that no spurious triggering was being caused by noise riding on the top or trailing edge of pulses which had already actuated the discriminator. Although in the above experiment the paralysis produced no noticeable effect in the measured counting rates, it was nevertheless retained as a precautionary measure.

counting rates other than 1 per second or amplifier time constants other than 3·2 microsec, the signal amplitude for a 50% count expressed in terms of the shot noise discriminator level, compared with the signal amplitude for a 50% count expressed in terms of the grid current noise discriminator level is varying in accordance

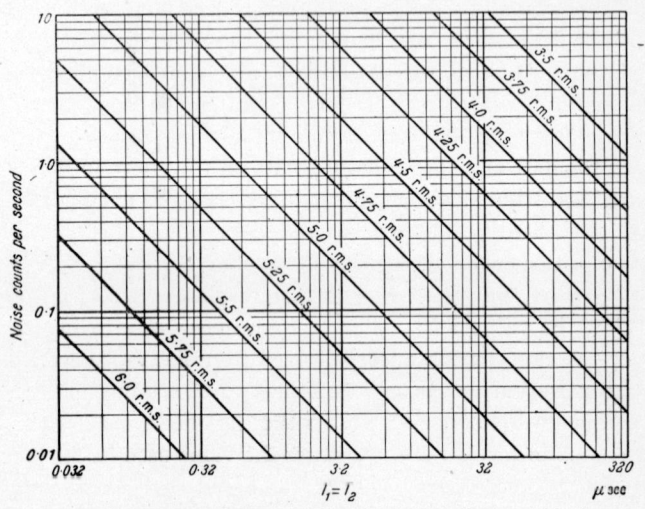

Figures on inclined lines indicate pulse amplitude in terms of r.m.s. noise level

Fig. 48. Pulse amplitude for a 50% count as a function of noise counting rate and equal amplifier time constants

with $1 - \dfrac{0 \cdot 09}{(z)^2}$ where z is the new grid current noise discriminator level relative to the 4·65 value appertaining to the grid current curve in fig. 47. This expression gives therefore a measure of the cancellation effect at any discriminator level. Considering then two x/v points near the right hand end of the curve in fig. 46, say 4·04 and 3·5, the ratio here is 1·16. The appropriate value for z is $\dfrac{3 \cdot 5}{4 \cdot 65} = 0 \cdot 75$ and substituting this in the experimental relationship above gives the figure 0·84. The product of this and the ratio 1·16 is 0·973 so the cancellation effect is still within 3%. Taking now two x/v points near the left hand end of the curve in fig. 46, say 6·0 and 5·66, the ratio this time is 1·06. The experimental relationship gives 0·94 and the product is 1·0. Hence the previous conditional conclusion that the same signal amplitude is required for both grid current noise and shot noise to give a 50% count can now be extended to cover any reasonable noise counting rate or amplifier bandwidth. This result is best illustrated graphically and is shown in fig. 48.

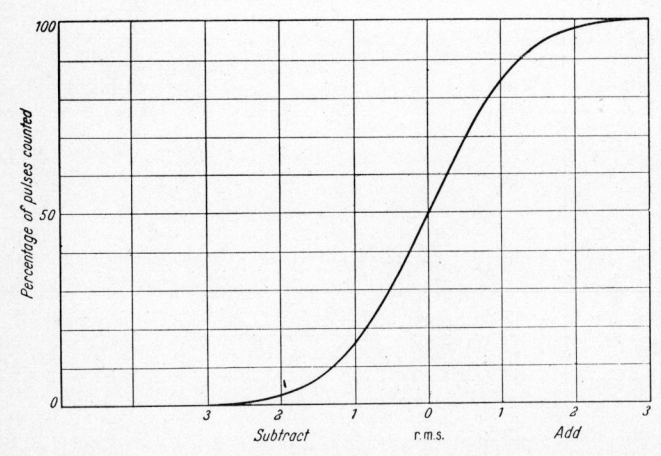

Fig. 49. Correction to pulse amplitude, for a count greater or less than 50%

For any required noise counting rate and for any amplifier bandwidth defined by the equal time constants T_1 and T_2, the figures on the inclined lines give the signal amplitude, relative to the r.m.s. noise, required in order that 50% of the signal pulses can be counted when the discriminator is set to give the chosen noise counting rate. To convert the percentage of signal pulses counted from 50% to any other desired figure the curve shown in fig. 49 is used.

This is simply a theoretical plot of the area under a Gaussian contour, since it has been satisfactorily checked earlier in the section that the signal counting rate does in fact follow this distribution. The r.m.s. figures have to be added to or subtracted from the value for a 50% count obtained from fig. 48.

To take an example, for a noise counting rate of 1 per second and for amplifier time constants each equal to 3·2 microsec, signal pulses of 4·65 times the r.m.s. noise level permit a 50% count to be obtained. For a 98% count the signal pulses then require an amplitude of $4·65 + 2·1 = 6·75$ times the r.m.s. noise. In all such calculations, the r.m.s. noise of the amplifier refers of course to the equivalent number of ion pairs at the input terminals as described in Section 5.1.

In the above example the curve of pulse rate against the discriminator level has the shape shown by the full line in fig. 50. (assuming the signal pulses are all the same size).

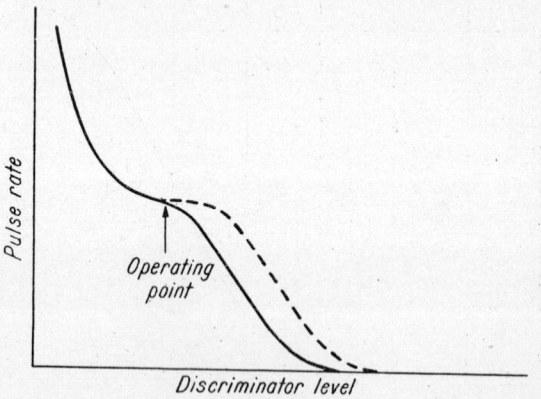

Fig. 50. Discriminator bias curves showing zero counting plateau and a plateau just equal to the r.m.s. noise

There is just one operating point to meet the imposed conditions and there is no safety margin against instrumental variations. Movement of the discriminator level above or below this point results in a loss in signal counts or a big increase in noise counts. The signal to noise ratio is in fact 1 and there is no workable plateau. The discriminator level at the operating point is very approximately 4·85 times the r.m.s. noise (actually, 4·65 for grid current noise and 5·07 for shot noise). If a working plateau equal to the r.m.s. noise is required, corresponding to a signal to noise ratio of $\frac{5\cdot85}{4\cdot85} = 1\cdot2$ and giving a $\pm$ 10% safety margin against instrumental variations, the signal pulses then require an amplitude of $4\cdot65 + 1 + 2\cdot1 = 7\cdot75$ times the r.m.s. noise. This is shown by the dotted curve in fig. 50 and is a typical example illustrating the use of the foregoing results in relation to the sensitivity of ionisation chamber amplifiers, used for total activity measurements.

It is not suggested that the analysis as developed is an exact one. It is expected however that it should be accurate to within 10%

in the large majority of cases and this is supported by experimental figures given in Table 3 at the end of this section. The interesting feature is of course its applicability to both grid current noise and shot noise. So far no mention has been made of the case where shot noise and grid current noise are present together and comparable in magnitude. It is difficult to visualise what will happen here and no simple analysis seems to be possible. It would not seem unreasonable however to assume that as the analysis is applicable to each type of noise separately, it probably applies to the mixture as well but with reduced accuracy. Some experimental figures are given later in the section and these would tend to support this suggestion.

It has already been explained how the sensitivity of an ionisation chamber and pulse amplifier used for total activity measurements can be measured by feeding in test pulses of known amplitude and recurrence frequency and measuring the proportion counted on a ratemeter. Since amplitude discriminators and ratemeters are easier come by and more robust than a thermal instrument reading r.m.s. volts, this method of measuring the sensitivity is largely used at A.E.R.E. In fact it is possible to put the analysis just described into reverse and to calculate from sensitivity measurements the r.m.s. noise level of the amplifier and where a quick estimation of the noise level is required, this is frequently done.

The following experimental figures in Table 3 show the sensitivities of amplifiers using EC91, 6AK5 and ME1400 input valves. The valves and input capacities were the same as those used for the noise level experiments in Section 5.1, the proportion of signal pulses counted was 98%* in each case and the permissible noise count was set at 1 per second. The corresponding figures calculated from the experimental noise level curves in Section 5.1 are included for comparison purposes.

For the time constant values where either shot noise or grid current noise predominates, 0·032 and 3·2 microsec. for the EC91, 0·08 and 8·0 microsec. for the 6AK5 and 3·2 and 32 microsec. for

---

* The figure of 98% was chosen because it was assumed that most experimentalists would be interested in counting all or nearly all of the signal pulses. For this particular experiment however, it was realised later on that a figure of 50% would have permitted higher accuracy in the results but as the input valves were no longer available it was not possible to do a repeat without extensive measurements first on the new valve parameters. This was not undertaken as the figures given in Table 3 are average values of a number of experiments and should therefore be reasonably accurate.

the ME1400, the agreement between the measured and calculated values is good, the error in each case not exceeding 10%. At 1·0 microsec. for the EC91 and 2·0 microsec. for the 6AK5 the grid current noise and shot noise components are approximately equal

TABLE 3

98% Signal Count.   1 Noise Count per Second.

Valve	$T_1 = T_2$	Sensitivity [Measured]	Sensitivity [Calculated]
	μs	Ion pairs	Ion pairs
E.C.91	0·032 1·0 3·2	26,000 8,300 10,000	24,600 7,350 10,200
6AK5	0·08 2·0 8·0	15,500 4,600 6,000	14,200 4,200 5,900
ME1400	3·2 32 320	8,300 2,800 2,200	8,600 2,620 1,850

and here the apparently larger error between the two sensitivity values would tend to support the suggestion that the analysis is applicable to the mixed noise case but with reduced accuracy. The largest measure of disagreement occurs with the ME1400 at 320 microsec. time constants and this is almost certainly due to the flicker noise component being comparable in magnitude with the other two. It would seem therefore, from the above experimental results, that the analysis, developed in this section, of the sensitivity of amplifiers used for total activity measurements is accurate enough for most practical purposes.

6

PROPORTIONAL AND SCINTILLATION COUNTERS.

In the following sections the main characteristics of proportional and scintillation counters are described. It is shown that the results which have been derived in the text with respect to ionisation chambers can be applied, with certain reservations, to materially assist in the solution of problems which tend to be peculiar to the use of these two types of counter.

6.1. PROPORTIONAL COUNTERS

The proportional counter, like the ion and electron chambers described in Chapter 2 relies on the ionisation produced in the filling gas for the detection and measurement of radiations from radioactive sources. Unlike the ion and electron chambers however, the proportional counter utilises a field intensity sufficiently high to impart to the electrons, in a mean free path length, enough energy to produce secondary electrons during subsequent collisions with neutral gas molecules. That is to say, the primary electrons are effectively multiplied by a factor M_g known as the gas multiplication and as a consequence the pulse from a proportional counter is many times greater than the pulse from an ionisation chamber. To realise a stable counter and also to achieve the high field intensity without excessive polarising voltage, the proportional counter usually takes the form of a cylindrical earthy electrode with a very fine wire along the axis forming the collecting electrode. This is shown in fig. 51.

With this type of geometry the electrostatic field inside the counter is radial and the field intensity at any radius r varies as $1/r \log b/a$ where a and b are the radii of the centre wire and earthy electrode respectively. The field intensity increases rapidly as the centre wire is approached and under normal working conditions the multiplication process takes place in a very small volume surrounding the wire. If the critical field intensity for production of secondary electrons occurs m mean free path lengths from the centre wire, the gas multiplication factor M_g is equal to 2^m. Multiplication factors between 10 and 10^4 are frequently used and under certain conditions it is possible to realise factors in excess of 10^6. The higher M_g values are however critically dependent on the counter polarising voltage and where good stability of the output pulse amplitude is required it is advisable to work with as low a gas multiplication as possible.

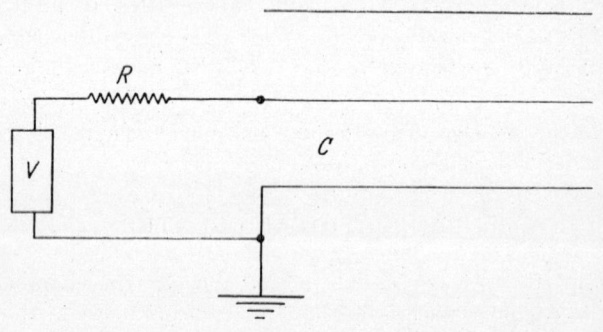

Fig. 51. Circuit arrangement of proportional counter

The most frequently used gas filling for the proportional counter is methane, but for multiplication factors not exceeding 10^2 argon is sometimes preferred. Argon is advantageous, in that for a given gas gain, it requires a lower polarising potential than methane.

Since the secondary ionisation is all produced in the very small volume surrounding the centre wire, the amplitude of the output pulse is virtually independent of the position of the primary ionisation in the counter and is of course proportional to the total primary ionisation produced. Hence the name proportional counter. The voltage pulse developed on the centre wire is due entirely to the motion of the positive ions away from the wire. The electrons, which are collected almost instantaneously and before the positive ions have begun to move, contribute nothing to the pulse because their charge is initially counterbalanced by the induced effect of the positive ions. As the ions drift outwards, the voltage pulse on the centre wire begins to rise, rapidly at first, as the ions are in a region of intense field and then more slowly as the ions move into the low field region near the earthy electrode. The pulse ultimately reaches an amplitude of $N_p e/C \times M_g$ if the time constant CR is long compared with the time taken by an ion to drift the full distance across the counter.

For primary ionisation localised inside the counter or distributed along a line parallel to the centre wire, the output pulse, expressed as a fraction of the final amplitude, varies with time according to[23].

$$\frac{1}{2 \log \frac{b}{a}} \log \left[\frac{b^2}{a^2} t + 1 \right].$$ This expression shows that the pulse reaches its maximum amplitude in a time proportional to $[1 - (a^2/b^2)]$ and reaches half amplitude in a time proportional to a/b. For a typical counter, such as is used at A.E.R.E.[24] with a centre wire of

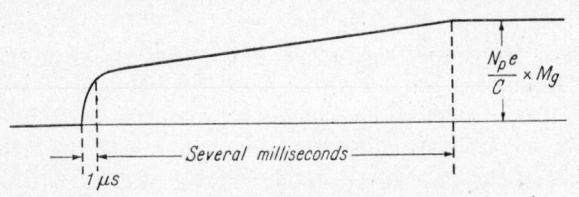

Fig. 52. Typical voltage pulse from proportional counter

1.2×10^{-3} cm diameter and an earthy electrode of 3.5 cm diameter, the pulse reaches half amplitude in a time of the order of 1 microsec. and reaches full amplitude in a time of the order of several milliseconds. The general shape of the proportional counter pulse is shown in fig. 52.

There is a fast initial rise up to about half amplitude, then a fairly rapid turn over and a much slower climb up to the maximum amplitude $\frac{N_p e}{C} \times M_g$. When the primary ionisation is distributed perpendicular to the centre wire, there is a time interval of some 2 microsec. during which the primary electrons are arriving at the wire. This has the effect of slightly slowing down and modifying the shape of the initial fast rise of the pulse, but produces no change in the subsequent slow climb, as in this region a time delay of 2 microsec. is negligible compared with the total time of several milliseconds to reach full amplitude.

In proportional counter work it is customary to differentiate the output pulse with a time constant which is very short compared with the total pulse duration of several milliseconds and which may in fact be comparable with the initial fast rise. This results in a pulse after differentiation whose amplitude is dependent entirely on the initial fast rise and whose width is very much less than the several milliseconds duration of the counter pulse. The resolving power of the system is thereby enormously improved and if certain precautions are taken (such as collimation of the tracks through the counter) the differentiated pulse amplitude can still be proportional to the energy of the ionising particle.

The collection time of a proportional counter is, in the strictest sense, the time taken for the pulse on the centre wire to reach its full amplitude, but in view of the short differentiation which is always used, it is customary when speaking of the collection time of a proportional counter to mean a time which is of the same order as the fast rising part of the pulse on the centre wire, i.e. 1 to 2 microsec. While it would appear impossible for the proportional counter to produce a pulse with a linear rise like that which can be obtained from an ionisation chamber, a proportional counter used primarily for activity measurements, that is, without collimation of particle tracks, does give an average shape of pulse with an initial rise which is considerably more linear than indicated by the expression $\dfrac{1}{2 \log \dfrac{b}{a}} \log \left[\dfrac{b^2}{a^2} t + 1 \right]$ This is, of course, due to the time delay effect mentioned above, and is confirmed experimentally in the next section.

It is suggested therefore, that for proportional counters which are used for assay work, and this probably covers the majority of present day applications, the results which have been derived in the text for ionisation chambers, particularly in regard to the effect of the amplifier differentiating and integrating time constants on the pulse amplitude and duration, can also be applied to proportional counters to assist in the solution of related problems.

6.2. EXPERIMENTAL DETERMINATION OF THE COLLECTION TIME OF A PROPORTIONAL COUNTER

The collection time of a proportional counter can be determined directly by feeding the counter pulse through a very wide band amplifier, in order not to distort the leading edge and by measuring the rise time on an oscilloscope[24]. The experiment described in this section was made to determine indirectly the collection time of a proportional counter by making use of the pulse amplitude curves in fig. 9, and to check in the process whether it is reasonable to apply such curves, computed for ionisation chambers, to proportional counters.

Basically the experiment measures the variation in pulse amplitude at the amplifier output for a known range of differentiating time constants and by comparing the experimental distribution with the theoretical curves it is possible to decide whether the use of such curves is justifiable and if so, to then proceed to a calculation of the collection time.

The amplifier used in the experiment had an upper frequency limit of 2 Mc/s (0·08 microsec. integration) but the integrating time constant was deliberately lengthened to 0·14 microsec. to ensure that the upper frequency fall off was really that of a single time constant. The differentiating time constants used were accurately determined by condenser and resistor measurements on a bridge before commencing the experiment and allowance was also made for additional capacity due to strays.

The proportional counter used was the A.E.R.E. model, with argon gas and operating at a polarising potential of 800 volts. The pulses were produced by an α particle source placed inside the counter and the count rate was measured directly by using a ratemeter after the amplitude discriminator. The discriminator trigger level was set to a point some distance above the normal counting plateau and as a consequence the ratemeter reading was very dependent on the mean amplitude of the larger pulses. Any change in the differentiating time constant which produced a simultaneous change in the mean pulse amplitude immediately showed up as an alteration in the ratemeter reading. The increase in amplifier gain necessary to bring the ratemeter reading back to its initial setting was noted for each value of the differentiating time constant, starting with the largest. The results are given in Table 4.

The figures for (Gain increase)$^{-1}$ relative to the 8·0 microsec. differentiating time constant effectively define the pulse amplitude distribution at the amplifier output. The next step is to fit this distribution (not the actual figures) as best one can to the amplitude curves in fig. 9, over the range of T_1/T_2 values in Table 4, endeavouring to make the experimental distribution lie along or parallel to a particular curve. In this case it so happens that the experimental distribution agrees closely with the theoretical curve for $T = 10T_2$.

TABLE 4

Diff. Time Constant μs	Ratio $\dfrac{T_1}{T_2}$	Gain increase relative to 8·0 μs D.T.C.	$\dfrac{1}{\text{Gain increase}}$
8·0	57	1	1
3·5	25	1·16	0·863
1·6	11·4	1·47	0·68
0·73	5·2	2·2	0·455
0·32	2·32	4·1	0·244
0·175	1·25	7·48	0·134
0·07	0·5	16·8	0·06

The experimental and theoretical points are compared in Table 5.

The agreement is remarkably good and since the amplifier time constants were carefully measured prior to the experiment, this

TABLE 5

Ratio $\dfrac{T_1}{T_2}$	Theoretical Points	Experimental Points
57	0·89	0·89
25	0·79	0·77
11·4	0·63	0·61
5·2	0·43	0·41
2·32	0·23	0·22
1·25	0·12	0·12
0·5	0·05	0·053

would indicate that the fast rising part of the counter pulse is more nearly linear than defined by the expression $\dfrac{1}{2 \log \dfrac{b}{a}} \log \left[\dfrac{b^2}{a^2} t + 1 \right]$.

This is presumably due to the majority of the α tracks being inclined to the centre wire. Thus it would appear that the curves in fig. 9 showing the effect of the amplifier bandwidth on a pulse having a constant rate of rise can be applied to proportional counters without incurring appreciable error.

The best fit is with the theoretical curve $T = 10\ T_2$ where T_2 is the amplifier fixed integrating time constant. In this case $T_2 = 0.14$

microsec. so this gives the average collection time of the proportional counter as 1·4 microsec.

6.3. PROPORTIONAL COUNTING IN THE PRESENCE OF AN INTENSE BACKGROUND OF WEAKLY IONISING PARTICLES

In experimental nuclear physics or in routine radiochemical assay, it is frequently required to count the signal pulses produced by relatively densely ionising particles in the unavoidable presence of pulses from other particles of a less densely ionising nature. Typical examples are the counting of fission fragments in the presence of α particles, α particles in the presence of β particles and protons in the presence of a high γ flux. This problem only becomes of importance when the background activity is sufficiently high to cause considerable overlapping with consequential fluctuating build up of the background pulses. If the background fluctuation required to simulate a wanted pulse is very large then the problem can be most easily solved by an application of the results derived in Section 5.2 for the counting of signal pulses in the presence of grid current noise. This condition usually holds in the case of α particles accompanied by an intense β background, and since this type of assay work is, at the time of writing, largely carried out by proportional counters, this particular problem has been singled out for special consideration in the present section. The general principles are however applicable to other pairs of particles and also to other types of counter.

When the background fluctuation required to simulate a wanted pulse is not very large, a condition which usually holds when counting fission pulses in the presence of an α background, other methods have to be adopted. This essentially more basic, yet if anything more difficult aspect of the problem is discussed in simplified form at the end of the present section.

Considering first the counting of α particles in the presence of a high β background. In a gas at N.T.P. α particles produce about 30,000 ion pairs per centimetre of track length (specific ionisation) while β particles on the average produce about 100 ion pairs per centimetre. Thus the α pulses passing through the amplifier are 300 times greater than the β pulses and it follows that unless the β activity is very high, there should be little difficulty in setting the discriminator level to record only the α pulses and to disregard the

β pulses. In some instances however, the β activity may be many thousands of times greater than the α activity and it is then no longer possible to ignore the probability of a large number of β pulses occurring during the response time of the amplifier and producing a fluctuation comparable in size with an α pulse. In fact the more frequent and useful presentation of the problem is to determine the maximum β activity which can be tolerated without the spurious counts due to the background exceeding some specified percentage of the true α counts. Put in this form, the particular merit of the problem is that it enables the possibility of success or failure of a proposed experiment to be estimated beforehand.

When the background fluctuation required to simulate an α pulse is very large (say 100 upwards) the background pulses behave in a manner similar to grid current noise and the amplitude distribution of the background becomes approximately Gaussian in nature. In such a case it is possible to make use of the results derived in Section 5.2 in relation to the counting of signals in the presence of grid current noise. It must be emphasised however, that the accuracy expected in the background problem will depend entirely on how valid is this assumption of a Gaussian distribution of amplitude and in the majority of practical cases will be considerably lower than the noise counting problem discussed in Section 5.2. In general however, an estimation of permissible background to within a factor of 2 is usually sufficient to decide prior to an experiment whether there is a reasonable chance of carrying it through to a successful conclusion.

Consider then the case of a proportional counter with a collection time of 1·5 microsec., which gives the full 300 to 1 ratio between the α and the β pulses. Suppose the amplifier time constants are both set to 32 microsec. and it is required to count not less than 98% of the α pulses while the β counts should not exceed 1 per second. The problem is to determine the maximum β activity which can be permitted. Since the amplifier time constants are long compared with the counter collection time T the latter can be neglected in this initial analysis. The curves in figs. 48 and 49 show that α pulses of amplitude not less than $4 \cdot 12 + 2 \cdot 1 = 6 \cdot 22$ times the r.m.s. β fluctuation are required to meet the stated conditions. The r.m.s. β fluctuation must therefore not exceed $\dfrac{300}{6 \cdot 22} = 48$ times the amplitude of a single β pulse. Now it was shown in Section 5.2 that the response time Δ of the system is the time interval during which the r.m.s. fluctuation in the mean number of pulses which occur, is just equal to the r.m.s. amplitude fluctuation expressed in terms of the maximum amplitude of a single pulse. Since the background pulses originate from step functions of voltage on the input grid of the amplifier, this case is analagous to grid current noise and the appropriate response time Δ is therefore $1 \cdot 85 \, T_1$ from equation (51).

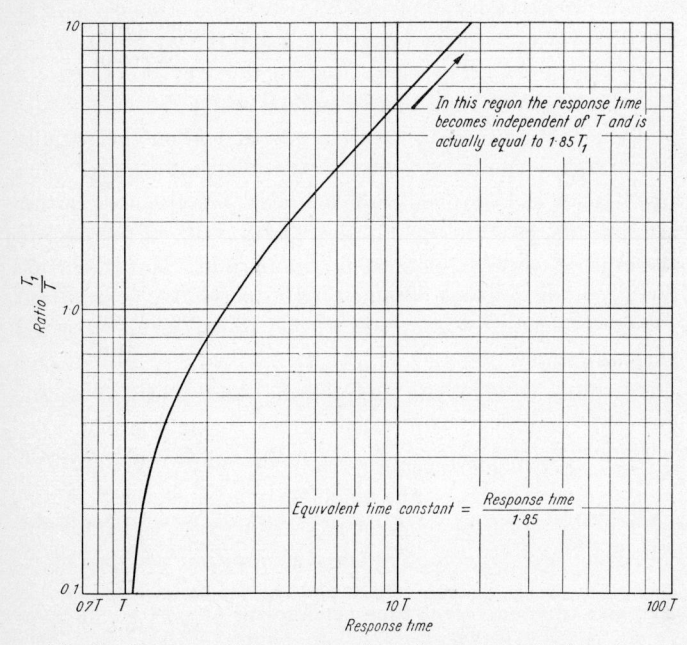

Fig. 53. Response time as a function of counter collection time and equal amplifier time constants

It follows that the maximum permissible β background to meet the stipulated conditions is given by $\dfrac{(48)^2}{\Delta} = \dfrac{(48)^2}{5 \cdot 9 \times 10^{-5}} = 3 \cdot 9 \times 10^7$ pulses per second. If a working plateau equal in length to the r.m.s. β fluctuation is required and giving a signal to noise ratio of $\dfrac{5 \cdot 12}{4 \cdot 12} = 1 \cdot 24$, the α pulses then must be at least 7·22 times the r.m.s. β fluctuation. This gives a maximum β background of $\left(\dfrac{300}{7 \cdot 22}\right)^2 \dfrac{1}{5 \cdot 9 \times 10^{-5}} = 2 \cdot 9 \times 10^7$ pulses per second.

From the foregoing example it is clear that the permissible β background can be increased substantially by decreasing the response time of the amplifier. This can be effected by decreasing the amplifier time constants T_1 and T_2 together until ultimately the response time is limited by the collection time of the counter. Beyond this point there is no improvement to be gained by further reducing T_1 and T_2. In the limit the response time is actually equal to the counter collection time T and the curve in fig. 53 shows how the response time varies with the ratio T_1/T when the latter is not large compared with unity. The curve was obtained by assuming that although the response time of the system is not the same as the resolving time they are both similarly affected when the amplifier time constants are comparable with or less than the counter collection time. When the time constants are large, fig. 31 shows that the response time of $1 \cdot 85\, T_1$ is equal to the resolving time if the discriminator level is set to 80% of the pulse amplitude. Thus the points for the curve in fig. 53 were obtained from the resolving time curves in fig. 31 for a constant discriminator level corresponding to 80% of the pulse amplitude.

For low values of T_1/T the curves in fig. 48 can still be applied provided the amplifier "equivalent time constant" defined as $\frac{\text{Response time}}{1 \cdot 85}$ is used in place of the actual value of T_1. For example, if in the previous calculation the amplifier time constants T_1 and T_2 are both reduced until they are equal to the counter collection time T, i.e. $1 \cdot 5$ microsec., fig. 53 gives the response time as $2 \cdot 35 T = 3 \cdot 5$ microsec. and the equivalent time constant as $\dfrac{3 \cdot 5}{1 \cdot 85} = 1 \cdot 9$ microsec. Using this equivalent value for T_1, the curves in figs. 48 and 49 give α pulses of $4 \cdot 75 + 2 \cdot 1 = 6 \cdot 85$ times the r.m.s. β fluctuation for zero plateau and $7 \cdot 85$ times for a plateau just equal to the r.m.s. fluctuation. In the former case therefore the maximum permissible β background is $\left(\dfrac{300}{6 \cdot 85}\right)^2 \dfrac{1}{3 \cdot 5 \times 10^{-6}} = 5 \cdot 5 \times 10^8$ pulses per second and in the latter case $4 \cdot 2 \times 10^8$ pulses per second.

The curve in fig. 53 is useful in that it shows, to a first order, the substantial improvement which can be obtained in the permissible β background by reducing the amplifier time constants, to equal the collection time of the counter. Further reduction in the time constants however results in a much slower increase in permissible background and in the limit when T_1 and T_2 are made very small, an improvement of slightly over 2 times is the most which can be realised on the above figures.

If the counter geometry is not ideal, as has been assumed so far in this analysis, but is such as to produce α and β pulses of varying amplitudes due to varying track lengths, it then becomes necessary to make the further assumption that the β fluctuation depends primarily on the mean amplitude of the β pulses, whereas the

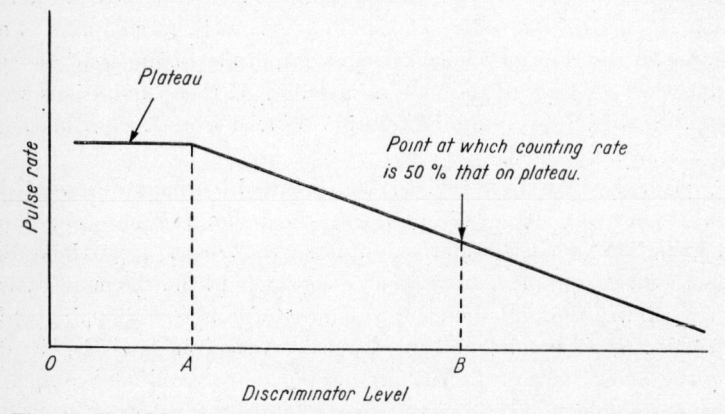

Fig. 54. Use of discriminator bias curve to determine average and minimum pulse amplitudes

minimum α pulse is the controlling amplitude if approximately 100% counting of the α pulses is required. These magnitudes can be obtained experimentally by measuring the counter plateau for a source containing α particles only. If this experimentally derived curve is as shown in fig. 54, the mimimum α pulse is approximately equal to OA while the average α pulse is approximately equal to OB. The average β pulse is therefore this latter magnitude divided by the ratio of the α to β specific ionisations, in this case 300.

For the proportional counter used at A.E.R.E. the average α pulse has an amplitude approximately 5 times greater than the minimum α pulse. This in turn corresponds to a ratio of 60 to 1 between the amplitudes of the minimum α pulse and the average β pulse. With this counter therefore the permissible β background for 1 count per second, amplifier time constants of 1·5 microsec. and no counting plateau is $\left(\dfrac{60}{6 \cdot 85}\right)^2 \dfrac{1}{3 \cdot 5 \times 10^{-6}} = 2 \cdot 2 \times 10^7$ pulses per second. The corresponding figure for a counting plateau just equal to the r.m.s. β fluctuation is 1.7×10^7 pulses per second. These background figures however must be taken with reserve and are undoubtedly much less accurate than the background figures previously calculated for the ideal geometry counter. This is only to be expected on account of the additional simplifying assumptions which have been made and also because of errors in estimating the minimum amplitude α pulse and the average amplitude β pulse from the counter characteristic. At the same time it is more than likely that the major part of the inaccuracy is due to the assumption of a Gaussian distribution of background amplitude for fluctuations as low as 60 β pulses to simulate an α pulse. If these limitations are appreciated however the information derived from the calculation is well worth having.

When the background fluctuation required to simulate a wanted pulse is not very large (say 20 or less) the assumption of a Gaussian distribution of background amplitude is no longer approximately valid and the problem must then be considered from the more basic viewpoint of the probability of a number of pulses overlapping and building up to simulate a wanted pulse. A rigorous analysis, based on the exact shape of the background pulses is exceedingly complex and it is doubtful if the answer when obtained is worthy of the effort involved. A much simpler, but also much less accurate approach is to replace each background pulse by an equivalent rectangular pulse, thereby transforming the problem into one which is more amenable to treatment using POISSON's formula. The equivalent rectangular pulse is usually assumed to have the same amplitude as the background pulse, while its duration can be considered as equal to the resolving time T_r given by equation (4) or alternatively the response time Δ given by equation (51). For equal time constants of differentiation and integration and T assumed small, equation (4) gives $T_r = 2.72\ T_1$ while from equation (51) $\Delta = 1.85 T_1$. It can be shown that rectangular pulses of duration $2.72 T_1$ give rise to the same mean fluctuating level as the true background pulses whereas those of duration $1.85 T_1$ give rise to the same r.m.s. fluctuation as the true background pulses. There would seem to be no obvious reason for choosing one duration in preference to the other. However, since the amplifiers under consideration do not pass the mean D.C. level of the background, it would appear that the response time Δ is probably the better choice. For this reason it is proposed to use Δ in the present analysis, and also because the curve in fig. 53 shows how Δ varies when the amplifier time constants are not long compared with the counter collection time.

Suppose $\bar{n}$ is the mean occurrence rate of the background pulses and H is the amplitude of the wanted pulses relative to the background. In an ideal geometry counter H would simply correspond to the ratio between the specific ionisations of the two particles. In a practical counter however, H would correspond to the ratio between the miminum amplitude wanted pulse and the average amplitude background pulse, as discussed earlier in this section. Now the mean number of background pulses occurring during the time of a single pulse is $\bar{n}\Delta$. For the background to simulate a wanted pulse, a fluctuation of magnitude H on the mean value $\bar{n}\Delta$ is required. This is the same thing as exactly $H + \bar{n}\Delta$ pulses occurring in a time interval Δ or alternatively $H + \bar{n}\Delta - 1$ pulses occurring in the time interval Δ after any one pulse. Thus the probability per pulse, p_H, of a fluctuation of magnitude H occurring is given by

$$p_H = \varepsilon^{-\bar{n}\Delta} \frac{(\bar{n}\Delta)^{H+\bar{n}\Delta-1}}{(H + \bar{n}\Delta - 1)!}$$

If $\bar{n}\Delta$ is much less than unity then the number of such fluctuations per second is simply $\bar{n}p_H$. If $\bar{n}\Delta$ is much greater than unity however, the exact relationship between the number of fluctuations per second and p_H is somewhat complex, but it has been shown [25] that an approximate expression valid for all values of $\bar{n}\Delta$ is $\dfrac{\bar{n}}{1 + \bar{n}\Delta} p_H$.

The author has compared the case for $\bar{n}\Delta$ very large with equation (48) giving the rate of counting of grid current noise crests and this comparison would indicate that $\dfrac{\bar{n}}{1 + \dfrac{\bar{n}\Delta}{3}} p_H$ is probably a better approximation. Furthermore, what is of interest in the practical case is not only those fluctuations which just simulate a wanted pulse, but also those which exceed this value since all such fluctuations will trigger the amplitude discriminator. Thus the number P_H of background fluctuations per second exceeding a trigger level H times the amplitude of a single background pulse is given by

$$P_H = \frac{\bar{n}}{1 + \dfrac{\bar{n}\Delta}{3}} \varepsilon^{-\bar{n}\Delta} \sum_{H}^{\infty} \frac{(\bar{n}\Delta)^{H+\bar{n}\Delta-1}}{(H + \bar{n}\Delta - 1)!} \qquad (54)$$

This expression is in such a form that P_H for a given $\bar{n}$ can be calculated directly but even so considerable labour is involved. When the permissible $\bar{n}$ for a given P_H is required, this can only be obtained by computing a curve showing $\bar{n}$ against P_H and deriving the answer from the curve. The following example will make this clear.

Preliminary experiments on a fission counter at A.E.R.E. showed that the minimum amplitude fission pulse corresponding to the top end of the counting plateau was approximately 17 times greater than the average amplitude α pulse, i.e. $H = 17$. It was then required to determine the maximum amount of fissile material which could be placed in the counter to give the highest possible fission count rate, preferably keeping the discriminator just on the counting plateau and without the α background exceeding 1 count per minute. The collection time of the counter was 0·1 microsec. and it was decided to use amplifier time constants of differentiation and integration also equal to 0·1 microsec. there being little advantage in going lower than this. The relevant value for Δ from fig. 53 is 0·23 microsec.

The curve of $\bar{n}$ against P_H computed from equation (54) is shown by the full line in fig. 55. This gives the permissible α background for 1 count per minute as approximately $1\cdot5 \times 10^7$ pulses per second. Experiment showed the correct figure to be nearer 10^7 pulses per second. Considering the many assumptions and approximations which have been made the agreement is not unreasonable. Further, since equation (54) makes no allowance for the reduction in amplitude suffered by the fission pulses due to background fluctuations of the opposite phase, there is here a tendency to overestimate the permissible value for $\bar{n}$, although this effect is only of consequence when $\bar{n}\Delta$ is appreciably greater than unity.

The dotted curve in fig. 55 shows the effect of making the drastic assumption of a Gaussian distribution of background amplitude and calculating the permissible background from the data in figs. 48 and 49. For the higher $\bar{n}\Delta$ values, where the assumption of a Gaussian distribution is at least tending to become more valid, the curve computed from equation (54) lies above that derived from figs. 48 and 49. This is in agreement with expectation, since it is in this region that the former curve is likely to overestimate the permissible value for $\bar{n}$. The lack of violent disagreement between the two curves for lower values of $\bar{n}\Delta$ would seem to indicate that for background

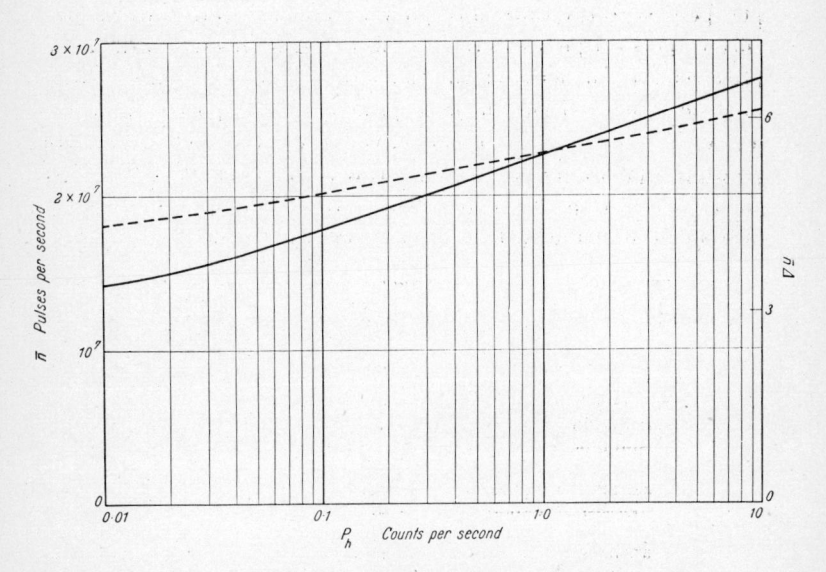

Fig. 55. α particle build up in a fission counter as a function of the amount of fissile material

fluctuations possibly as low as 15 times the amplitude of a single pulse, a figure which just encompasses the well designed fission counter, a quick approximation of the permissible background activity can be obtained from the data in figs. 48 and 49, thereby avoiding the labour associated with equation (54).

As a general summing up to the background problem therefore, it can be said that where very large fluctuations are involved some fair measure of reliance can be placed on the results computed from figs. 48 and 49. Where the fluctuations are small however, the results should be taken with great reserve and regarded more as giving initial guidance, the final answer having to await experimental verification.

6.4. SCINTILLATION COUNTERS

The scintillation counter is not a gas ionisation device like the other counters previously described, but relies on light scintillations produced in a phosphor for the detection and measurement of radiations from radioactive sources. The energy expended in the phosphor is transformed into photons which in turn give rise to a current pulse of photoelectrons from the light sensitive cathode of a photocell. In modern scintillation counters it is customary to use multiplier photocells in which the primary current pulse is multiplied many times by a series of successive stages of secondary emission. Multiplier photocells in use at A.E.R.E. have 11 stages and give gains varying from 10^6 to 10^8. This corresponds to an average gain of from 3·5 to 5·5 in each stage of secondary emission.

TABLE 6

Phosphor	Chief Application	Scintillation Time μS
Nickel-killed zinc sulphide	α particles	3.0
Thallium-activated potassium iodide	$\alpha\ \beta$ particles γ rays	1.5
Napthalene + 0.1% anthracene	β particles γ rays	0.06
Anthracene	β particles γ rays	0.02

Although the photomultiplier is essentially an electronic device, it is customary to consider a scintillation counter as a unit comprising the phosphor, the optical system and the photomultiplier itself. The gain required in any subsequent valve amplifier is usually very low and if photomultipliers having higher gains and output currents are produced in the near future, the need for an amplifier will probably not arise and a simple cathode follower used primarily as an impedance transforming device should suffice.

The light flash produced in the phosphor of a scintillation counter is usually assumed to die away exponentially with a time constant which is a characteristic of the phosphor material. This time constant is known as the scintillation time of the phosphor and may vary from a very small fraction of a microsecond to several microseconds. Again, some phosphors are found to be more efficient for one type of radiation than for others and it is therefore customary to characterise a particular phosphor by specifying the radiation for which it is most suitable. Table 6 lists a few of the phosphors

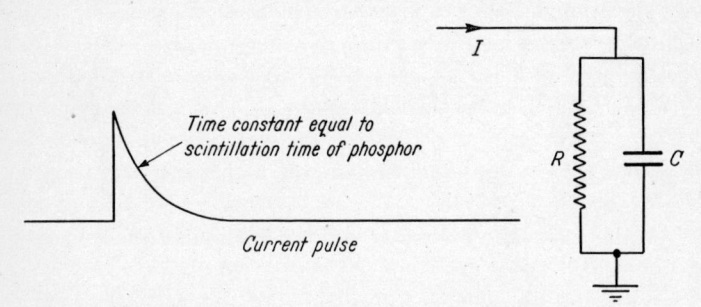

Fig. 56. Typical signal current pulse from scintillation counter and equivalent circuit in final anode of photomultiplier

commonly used at A.E.R.E. together with their chief applications and scintillation times.

The current pulse at the photomultiplier final anode has the same shape as the light flash in the phosphor, that is, it rises practically instantaneously to its maximum value and then decays exponentially with a time constant equal to the scintillation time of the phosphor. The final anode of the photomultiplier is fed through a resistor R

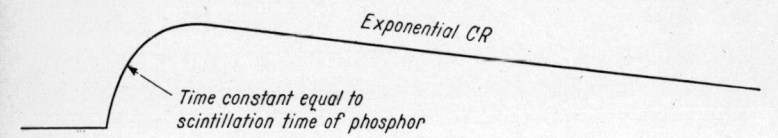

Fig. 57. Typical signal voltage pulse from scintillation counter

and the output voltage pulse is produced by the current pulse flowing into the circuit comprising R shunted by the stray capacity C of the photomultiplier. This is shown in fig. 56.

If it is assumed that the anode circuit time constant CR is very long compared with the phosphor scintillation time, the current pulse produces across the condenser C a voltage pulse having an exponential rise with a time constant equal to the scintillation time of the phosphor. Thereafter the voltage pulse slowly falls to zero with the anode circuit time constant CR. This is shown in fig. 57.

To secure high resolving power from the scintillation counter it is customary to differentiate the output pulse shown in fig. 57 with a time constant of the same order as the scintillation time of the phosphor. Since the valve amplifier used with a scintillation counter has a very low gain and there is consequently complete freedom from valve and circuit noise, the differentiating time constant in this case need not be placed within the amplifier subsequent to the first valve but can in fact be the anode circuit time constant of the photomultiplier. For practical reasons which tend to prohibit the use of very high anode loads this course is nearly always followed and thus CR is usually kept small and comparable with the scintillation time of the phosphor.

Since the scintillation counter is not a gas ionisation device, it is not possible to talk about its collection time. Nevertheless, the reader will not be slow to appreciate the similarity between the scintillation counter pulse shown in fig. 57 and the pulse from a gas ionisation counter, particularly if the scintillation time of the phosphor is regarded as being analogous to the collection time of the gas counter. Although the scintillation counter pulse has an exponential leading edge and not a linear one as has been assumed and preferred throughout the text, it is of interest to enquire whether, for computation purposes, the exponential pulse might not be approximately replaced by a pulse having a linear rise, not necessarily occurring in a time equal to the scintillation time of the phosphor. To check this the following simple experiment was carried out.

Artificial pulses, similar to that shown in fig. 57 and having an exponential leading edge of 3·0 microsec. time constant were fed through the amplifier which was used for the proportional counter experiment in Section 6.2 and the variation in pulse amplitude at the amplifier output was measured for a range of differentiating time constants. Since for a given differentiating time constant, the pulses at the amplifier output were all the same amplitude, it was a relatively simple matter to measure this amplitude using only a discriminator, without the additional complexity of a ratemeter as in the previous experiment. The integrating time constant was this time kept fixed at 0·8 microsec. and the two shortest differentiating time constants were changed to the longer values of 15·5 microsec. and 38·5 microsec.

The object of the experiment was again to compare the pulse

amplitude distribution at the amplifier output with the theoretical distribution curves in fig. 9 for pulses having a constant rate of rise and to determine from this comparison the best substitute for the exponential type of pulse fed into the amplifier. It was found that the amplitude distribution at the amplifier output agreed reasonably well with the theoretical curve for $T = 5T_2$. The comparison between the theoretical and experimental points is given in Table 7.

TABLE 7

Ratio $\dfrac{T_1}{T_2}$	Theoretical Points	Experimental Points
48	0.91	0.92
19.5	0.81	0.81
10	0.71	0.67
4.4	0.53	0.49
2.0	0.34	0.33
0.92	0.18	0.2
0.4	0.08	0.1

The curve defined by $T = 5T_2$ here corresponds to a pulse having a leading edge rising linearly in a time of $5 \times 0.8 = 4$ microsec., which is some 30% greater than the time constant of the applied exponential pulse. It follows, that for many applications, particularly where the effect of the amplifier bandwidth on a pulse from a scintillation counter is under consideration, the latter can be replaced by a pulse having a linear rise in a time some 30% greater than the scintillation time of the phosphor and that the relevant conclusions derived in the text for ionisation chambers are then applicable to the scintillation counter without producing excessive error.

6.5. SCINTILLATION COUNTER BANDWIDTH FOR OPTIMUM SIGNAL TO NOISE RATIO

The major source of noise in a scintillation counter is the photomultiplier tube itself. Valve and circuit noise produced in the subsequent electronic amplifier is so low compared with the signal level that it can be completely neglected. The photomultiplier noise arises primarily from the spontaneous thermal emission of single electrons from the photo cathode and the first few dynodes, each electron being multiplied by the subsequent stages of secondary emission and giving rise to small voltage pulses in the anode circuit. These noise pulses are not all the same size but vary above and below an average value due to the statistical variation in the multiplication factor of each stage of secondary emission. Occasional noise pulses rise to a height of some 8 times that of the average pulse[26]. The noise pulses in the anode circuit, prior to differentiation have the general shape as shown in fig. 58.

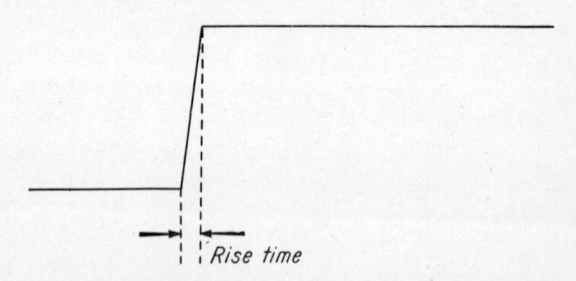

Fig. 58. Typical noise voltage pulse from scintillation counter

The rise time is primarily determined by the spread in the transit time of a single electron passing through all the stages of secondary emission and is usually of the order of 5×10^{-9} sec.

The noise at the amplifier output is due to the random summation of a large number of pulses, such as in fig. 58, after passing through the system time constants of differentiation and integration. If these time constants, assuming them to be equal, are initially long, there is considerable overlapping of the noise pulses at the amplifier output and appreciable fluctuating build up occurs. Decreasing the time constants together does not affect the amplitudes of the individual noise pulses but the fluctuating build up amplitude decreases due to the noise pulse duration shortening in step with the amplifier time constants. When the time constants are lower than the mean interval between noise pulses, the build up is relatively small but there is still a reasonable probability of 2, even 3 pulses occasionally overlapping. For an average tube the total number of noise pulses per second is usually in the region of 5×10^4, giving a mean interval between pulses of 20 microsec. Thus for time constants of say 2 microsec., corresponding to a response time Δ of

3·7 microsec., there is overlapping of 2 noise pulses approximately 9000 times per second and overlapping of 3 noise pulses approximately 900 times per second. For time constants of 0·2 microsec. the corresponding figures are 900 and 9 per second. Consequently the noise level is improved by decreasing the system time constants of differentiation and integration, rapidly at first when appreciable build up occurs and then more slowly when the resolution becomes sufficient to separate out the individual noise pulses.

The signal pulse differs from the noise pulses in that the leading edge, prior to differentiation is exponential in form with a time constant equal to the scintillation time of the phosphor. As was shown in Section 6.4 this pulse can be considered as approximately equivalent to a pulse having a linear rise in a time some 30% greater than the scintillation time of the phosphor. The amplitude of the signal pulse at the amplifier output is unaffected by a reduction in the time constants of differentiation and integration until the latter are approximately equal to the rise time of the signal pulse (see fig. 9). During this stage the signal to noise ratio steadily improves. Further reduction of the system time constants causes the signal pulse to fall in amplitude, slowly at first and then more rapidly resulting in the signal to noise ratio passing through an optimum. As far as can be deduced from this simple analysis therefore, it would appear that the system time constants of differentiation and integration should be reduced until equal to, or perhaps just slightly greater than the scintillation time of the phosphor. This result is applicable to all cases where the scintillation time of the phosphor is longer than the rise time of the noise pulses and covers the majority of phosphors in present day use. Even for a phosphor with a scintillation time as short as 0·02 microsec. the above result only demands an amplifier with an upper half power frequency of 8 Mc/s. Bearing in mind the low gain which is required this is well within the capabilities of conventional high slope receiving valves.

There are many other factors which affect the signal to noise ratio of a scintillation counter, the most important being statistical fluctuations in the number of signal photoelectrons emitted from the photo cathode[26], [27]. Also, it has been shown that the signal to noise ratio can be improved by cooling the photomultiplier with dry ice or liquid nitrogen[26] to reduce the emission rate of the thermal electrons and again, systems such as the use of two scintillation counters in coincidence[28] can also give substantial improvements in the signal to noise ratio. As far as judicial choice of the amplifier bandwidth is concerned however, the foregoing conclusions are perfectly general and are not invalidated in any way by the additional factors just mentioned.

REFERENCES

[1] JOHNSON, E. A. and JOHNSON, A. G.; A Theoretical Analysis of the Operation of Ionisation Chambers and Pulse Amplifiers. Phys. Rev. (July) 1936 **50** 170.

[2] GILLESPIE, A. B.; An Introduction to A.C. Amplifiers for Nuclear Physics Research. A.E.R.E. Report G/R 118, August 1947.

[3] GILLESPIE, A. B.; The Design of A.C. Amplifiers for use with Ionisation Chambers. A.E.R.E. Report G/R 168, February 1948.

[4] ELMORE, W. C.; Electronics for the Nuclear Physicist. Nucleonics (March) 1948 **2** 16.

[5] CURRAN, S. C. and CRAGGS, J. D.; Counting Tubes. Butterworths Scientific Publications, London 1949.

[6] WILSON, R.; Noise in Ionisation Chamber Amplifiers. Phil. Mag. (Jan.) 1950 **41** 66.

[7] KORFF, S. A.; Electron and Nuclear Counters. D. Van Nostrand Co. Inc., New York 1946.

[8] FRISCH, O. R.; Isotope Analysis of Uranium Samples by means of their Alpha-ray Groups. British Atomic Energy Project BR 49 1944.

[9] MOULLIN, E. B.; Spontaneous Fluctuations of Voltage. Clarendon Press, Oxford 1938.

[10] GOLDMAN, S.; Frequency Analysis Modulation and Noise. McGraw Hill Book Co. Inc., New York 1948.

[11] JOHNSON, J. B.; Thermal Agitation of Electricity in Conductors. Phys. Rev. (July) 1928 **32** 97.

[12] NYQUIST, H.; Thermal Agitation of Electric Charge in Conductors. Phys. Rev. (July) 1928 **32** 110.

[13] SCHOTTKY, W.; Spontaneous Current Fluctuations in Various Conductors. Ann. Physik. 1918 **57** 541.

[14] NORTH, D. O.; Fluctuation Noise in Space Charge Limited Currents at Moderately High Frequencies, R.C.A. Review (April) 1940 **4** 441; (July) 1940 **5** 106; (Oct.) 1940 **5** 244.

[15] JOHNSON, J. B.; The Schottky Effect in Low Frequency Circuits. Phys. Rev. (July) 1925 **26** 71.

[16] HARRIS, E. J.; Noise Data for Low Audio Frequency Apparatus. Paper II of C.V.D. discussion on Flicker Noise, July 1947, Dept. of Physical Research, Admiralty.

[17] COOKE-YARBOROUGH, E. H., BRADWELL, J., FLORIDA, C. D. and HOWELLS, G. A.; A Pulse Amplitude Analyser of Improved Design. Proc. Inst. Elect. Engrs. (Part III) 1950 **97** 108.

[18] BUNEMAN, O., CRANSHAW, T. E. and HARVEY, J. A.; The Design of Grid Ionisation Chambers. Can. J. Res. 1949 **27** 191.

[19] CAMPBELL, N. R.; The Study of Discontinuous Phenomena. Proc. Camb. Phil. Soc. 1908 **15** 117.

[20] CAMPBELL, N. R. and FRANCIS, V. J.; A Theory of Valve and Circuit Noise. J. Inst. Elect. Engrs. (Part III) 1946 **93** 45.
[21] CAMPBELL, N. R. and FRANCIS, V. J.; Random Fluctuations in a Cathode Ray Oscillograph. Phil. Mag. 1946 **37** 289.
[22] COOKE-YARBOROUGH, E. H. and PULSFORD, E. W.; A Counting Rate Meter of High Accuracy. Proc. Inst. Elect. Engrs. (Part II) 1951 **98** 191.
[23] WILKINSON, D. H.; Ionisation Chambers and Counters. Cambridge University Press 1950.
[24] SHARPE, J. and TAYLOR, D.; Nuclear Particle and Radiation Detectors. Part 2, Counters and Counting Systems. Proc. Inst. Elect. Engrs. (Part II) 1951 **98** 209.
[25] ROSSI, B. and STAUB, H.; Ionisation Chambers and Counters. McGraw Hill Book Co. Inc., New York 1949.
[26] MORTON, G. A. and MITCHELL, J. A.; Performance of a 931-A Type Multiplier as a Scintillation Counter. Nucleonics (Jan.) 1949 **4** 16.
[27] HOYT, ROSALIE C.; The Scintillation Counter as a Proportional Device. Rev. Sci. Inst. 1949 **20** 178.
[28] MORTON, G. A. and ROBINSON, K. W.; A Coincidence Scintillation Counter. Nucleonics (Feb.) 1949 **4** 25.

APPENDICES

APPENDIX A

Consider first the input signal (fig. 8) passing through the time constant of differentiation $C_1 R_1 = T_1$. Let V_{in} represent the input voltage and V_R the output voltage across R_1 at any instant t. The solution consists of two parts, one valid for time 0 to T and the other valid for time T to ∞.

Time 0 to T

The differential equation expressing the circuit conditions is

$$V_R = V_{in} - \frac{1}{C_1} \int \frac{V_R}{R_1} dt$$

Differentiating gives

$$\frac{dV_R}{dt} = \frac{dV_{in}}{dt} - \frac{V_R}{C_1 R_1}$$

$$= \frac{V_0}{T} - \frac{V_R}{T_1}$$

Multiplying both sides by ε^{t/T_1} gives

$$\frac{d}{dt}(V_R \varepsilon^{t/T_1}) = \frac{V_0}{T} \varepsilon^{t/T_1}$$

$$\therefore V_R \varepsilon^{t/T_1} = \frac{V_0 T_1}{T} \varepsilon^{t/T_1} + K_1$$

where K_1 is a constant of integration.

At time $t = 0$ $V_R = 0$ hence by substitution

$$K_1 = -\frac{V_0 T_1}{T}$$

$$\therefore V_R = \frac{V_0 T_1}{T} - \frac{V_0 T_1}{T}(\varepsilon^{-t/T_1})$$

$$= \frac{V_0 T_1}{T}\left[1 - \varepsilon^{-t/T_1}\right] \qquad \text{(a)}$$

Time T to ∞

The differential equation is as before

$$V_R = V_{in} - \frac{1}{C_1}\int \frac{V_R}{R_1}\,dt$$

Differentiating this time gives

$$\frac{dV_R}{dt} = -\frac{V_R}{T_1} \text{ as } V_{in} \text{ is now constant.}$$

Multiplying by ε^{t/T_1} gives

$$\frac{d}{dt}(V_R\,\varepsilon^{t/T_1}) = 0$$

$$\therefore V_R\,\varepsilon^{t/T_1} = K_2$$

At time $t = T$ $V_R = \dfrac{V_0 T_1}{T}\left[1 - \varepsilon^{-T/T_1}\right]$ given by equation (a) above. Hence by substitution

$$K_2 = \frac{V_0 T_1}{T}\left[1 - \varepsilon^{-T/T_1}\right]\varepsilon^{T/T_1}$$

$$\therefore V_R = \frac{V_0 T_1}{T}\left[\varepsilon^{T/T_1} - 1\right]\varepsilon^{-t/T_1} \qquad (b)$$

Hence the signal after passing through the time constant of differentiation T_1 is defined by equations (a) and (b) above and has the shape as shown in fig. A.1.

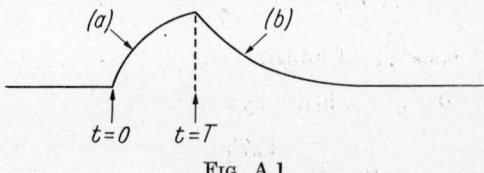

Fig. A.1

Consider now this signal passing through the time constant of integration $C_2 R_2 = T_2$. Let V_{in} represent the input voltage and V_c the output voltage across C_2 at any instant t. As before the solution consists of two parts, one valid for time 0 to T and the other valid for time T to ∞.

Time 0 to T

The differential equation is

$$V_c = V_{in} - C_2 R_2 \frac{dV_c}{dt}$$

$$\therefore \frac{dV_c}{dt} + \frac{V_c}{T_2} = \frac{V_{in}}{T_2}$$

$$= \frac{V_0 T_1}{T\, T_2}\left[1 - \varepsilon^{-t/T_1}\right]$$

since V_{in} is given in this case by the expression in equation (a)

$$\therefore \frac{d}{dt}(V_c \varepsilon^{t/T_2}) = \frac{V_0 T_1}{T\, T_2}\left[\varepsilon^{t/T_2} - \varepsilon^{t\left(\frac{T_1-T_2}{T_1 T_2}\right)}\right]$$

and $\quad V_c\, \varepsilon^{t/T_2} = \dfrac{V_0 T_1}{T\, T_2}\left[T_2 \varepsilon^{t/T_2} - \left(\dfrac{T_1 T_2}{T_1 - T_2}\right)\varepsilon^{t\left(\frac{T_1-T_2}{T_1 T_2}\right)}\right] + K_3$

When $t = 0$ $V_c = 0$ therefore by substitution

$$K_3 = -\frac{V_0 T_1}{T\, T_2}\left[T_2 - \frac{T_1 T_2}{T_1 - T_2}\right]$$

Replacing V_c by $V_{(0-T)}$ in keeping with the terminology used in the text gives

$$V_{(0-T)} = \frac{V_0 T_1}{T}\left[1 - \varepsilon^{-t/T_2}\right] - \frac{V_0 T_1^2}{T(T_1 - T_2)}\left[\varepsilon^{-t/T_1} - \varepsilon^{-t/T_2}\right] \quad (1)$$

Time T to ∞

The differential equation is

$$V_c + C_2 R_2 \frac{dV_c}{dt} = V_{in}$$

$$= \frac{V_0 T_1}{T}\left[\varepsilon^{T/T_1} - 1\right]\varepsilon^{-t/T_1}$$

as V_{in} this time is given by equation (b)

$$\therefore \frac{dV_c}{dt} + \frac{V_c}{T_2} = \frac{V_0 T_1}{T\, T_2}\left[\varepsilon^{T/T_1} - 1\right]\varepsilon^{-t/T_1}$$

$$\therefore \frac{d}{dt}(V_c\, \varepsilon^{t/T_2}) = \frac{V_0 T_1}{T\, T_2}\left[\varepsilon^{T/T_1} - 1\right]\varepsilon^{t\left(\frac{T_1-T_2}{T_1 T_2}\right)}$$

$$\therefore V_c \varepsilon^{t/T_2} = \frac{V_0 T_1}{T\, T_2}\left[\varepsilon^{T/T_1}-1\right]\frac{T_1 T_2}{T_1-T_2}\varepsilon^{t\left(\frac{T_1-T_2}{T_1 T_2}\right)} + K_4$$

Now when $t = T$

$$V_c = \frac{V_0 T_1}{T}\left[1 - \varepsilon^{-T/T_2}\right] - \frac{V_0 T_1^2}{T(T_1-T_2)}\left[\varepsilon^{-T/T_1} - \varepsilon^{-T/T_2}\right]$$

given by equation (1). Therefore by substitution

$$K_4 = \frac{V_0 T_1}{T}\left[\varepsilon^{T/T_2}-1\right] - \frac{V_0 T_1^2}{T(T_1-T_2)}\left[\varepsilon^{T\left(\frac{T_1-T_2}{T_1 T_2}\right)}-1\right]$$
$$- \frac{V_0 T_1^2}{T(T_1-T_2)}\left[\varepsilon^{T/T_2} - \varepsilon^{T\left(\frac{T_1-T_2}{T_1 T_2}\right)}\right]$$
$$= \frac{V_0 T_1}{T}\left[\varepsilon^{T/T_2}-1\right] - \frac{V_0 T_1^2}{T(T_1-T_2)}\left[\varepsilon^{T/T_2}-1\right]$$
$$= -\frac{V_0 T_1 T_2}{T(T_1-T_2)}\left[\varepsilon^{T/T_2}-1\right]$$

Hence

$$V_c \varepsilon^{t/T_2} = \frac{V_0 T_1^2}{T(T_1-T_2)}\left[\varepsilon^{T/T_1}-1\right]\varepsilon^{t\left(\frac{T_1-T_2}{T_1 T_2}\right)}$$
$$- \frac{V_0 T_1 T_2}{T(T_1-T_2)}\left[\varepsilon^{T/T_2}-1\right]$$

and replacing V_c by $V_{(T-\infty)}$ gives

$$V_{(T-\infty)} = \frac{V_0 T_1^2}{T(T_1-T_2)}\left[\varepsilon^{T/T_1}-1\right]\varepsilon^{-t/T_1}$$
$$- \frac{V_0 T_1 T_2}{T(T_1-T_2)}\left[\varepsilon^{T/T_2}-1\right]\varepsilon^{-t/T_2}. \quad (2)$$

Equations (1) and (2) define completely the pulse waveform after the operations of differentiation and integration, that is, the pulse reaching BB' in the equivalent circuit. It is now required to determine the maximum amplitude of this pulse compared with the amplitude V_0 of the input pulse. A physical consideration of the problem shows that the maximum does not occur during the time 0 to T. For instance, during this time the signal passing the differentiating time constant is defined by equation (a) and the form

APPENDIX A

of this expression is an exponential build up which always has a positive rate of change of voltage. This exponential build up is the input signal to the time constant of integration and since the output voltage which appears across C_2 can never exceed the input while the latter is increasing, the charging current through R_2 must be unidirectional and hence the rate of change of output voltage is always positive. Accordingly the pulse maximum must occur during the time T to ∞ and the relevant expression is given by equation (2).

Differentiating the latter with respect to time and equating to zero gives,

$$0 = -\frac{V_0 T_1}{T(T_1 - T_2)}\left[\varepsilon^{T/T_1} - 1\right]\varepsilon^{-t/T_1}$$

$$+ \frac{V_0 T_1}{T(T_1 - T_2)}\left[\varepsilon^{T/T_2} - 1\right]\varepsilon^{-t/T_2}$$

$$\therefore \left[\varepsilon^{T/T_1} - 1\right]\varepsilon^{-t/T_1} = \left[\varepsilon^{T/T_2} - 1\right]\varepsilon^{-t/T_2}$$

Taking logs to base ε on both sides gives

$$-\frac{t}{T_1} + \log_\varepsilon\left[\varepsilon^{T/T_1} - 1\right] = -\frac{t}{T_2} + \log_\varepsilon\left[\varepsilon^{T/T_2} - 1\right]$$

Therefore the time t_m at which the maximum occurs is given by

$$t_m = \frac{T_1 T_2}{T_1 - T_2} \log_\varepsilon \frac{\varepsilon^{T/T_2} - 1}{\varepsilon^{T/T_1} - 1} \tag{c}$$

To find the maximum value V_m substitute this value for t_m in equation (2) remembering that $\varepsilon^{\log_\varepsilon a} = a$ and $a \log x = \log (x)^a$. This gives

$$V_m = \frac{V_0 T_1^2}{T(T_1 - T_2)} \frac{\left[\varepsilon^{T/T_1} - 1\right]^{\frac{T_1}{T_1 - T_2}}}{\left[\varepsilon^{T/T_2} - 1\right]^{\frac{T_2}{T_1 - T_2}}} - \frac{V_0 T_1 T_2}{T(T_1 - T_2)} \frac{\left[\varepsilon^{T/T_1} - 1\right]^{\frac{T_1}{T_1 - T_2}}}{\left[\varepsilon^{T/T_2} - 1\right]^{\frac{T_2}{T_1 - T_2}}}$$

$$= \frac{V_0 T_1}{T} \frac{\left[\varepsilon^{T/T_1} - 1\right]^{\frac{T_1}{T_1 - T_2}}}{\left[\varepsilon^{T/T_2} - 1\right]^{\frac{T_2}{T_1 - T_2}}} \tag{3}$$

APPENDIX B

Evaluation of the area under the pulse.

$$V_{(0-T)} = \frac{V_0 T_1}{T}\left[1 - \varepsilon^{-t/T_2}\right] - \frac{V_0 T_1^2}{T(T_1 - T_2)}\left[\varepsilon^{-t/T_1} - \varepsilon^{-t/T^2}\right] \quad (1)$$

$$= \frac{V_0 T_1}{T}\left[1 - \varepsilon^{-t/T_2} - \frac{T_1}{T_1 - T_2}\varepsilon^{-t/T_1} + \frac{T_1}{T_1 - T_2}\varepsilon^{-t/T_2}\right]$$

$$= \frac{V_0 T_1}{T}\left[1 - \frac{T_1}{T_1 - T_2}\varepsilon^{-t/T_1} + \frac{T_2}{T_1 - T_2}\varepsilon^{-t/T_2}\right]$$

$$\therefore \int_0^T V_{(0-T)} = \frac{V_0 T_1}{T}\left[t + \frac{T_1^2}{T_1 - T_2}\varepsilon^{-t/T^1} - \frac{T_2^2}{T_1 - T_2}\varepsilon^{-t/T^2}\right]_0^T$$

$$= \frac{V_0 T_1}{T}\left[T + \frac{T_1^2}{T_1 - T_2}\varepsilon^{-T/T_1} - \frac{T_2^2}{T_1 - T_2}\varepsilon^{-T/T_2}\right.$$

$$\left. - \frac{T_1^2}{T_1 - T_2} + \frac{T_2^2}{T_1 - T_2}\right]$$

$$= \frac{V_0 T_1}{T}\left[T - T_1 - T_2 + \frac{T_1^2}{T_1 - T_2}\varepsilon^{-T/T_1}\right.$$

$$\left. - \frac{T_2^2}{T_1 - T_2}\varepsilon^{-T/T_2}\right]$$

Similarly

$$V_{(T-\infty)} = \frac{V_0 T_1^2}{T(T_1 - T_2)}\left[\varepsilon^{T/T_1} - 1\right]\varepsilon^{-t/T_1} - \frac{V_0 T_1 T_2}{T(T_1 - T_2)}$$

$$\left[\varepsilon^{T/T_2} - 1\right]\varepsilon^{-t/T_2} \quad (2)$$

$$= \frac{V_0 T_1}{T}\left[\frac{T_1}{T_1 - T_2}(\varepsilon^{T/T_1} - 1)\varepsilon^{-t/T_1} - \frac{T_2}{T_1 - T_2}\right.$$

$$\left.(\varepsilon^{T/T_2} - 1)\varepsilon^{-t/T_2}\right]$$

APPENDIX C

$$\therefore \int_T^\infty V_{(T-\infty)} = \frac{V_0 T_1}{T} \left[\frac{-T_1^2}{T_1 - T_2} (\varepsilon^{T/T_1} - 1)\varepsilon^{-t/T_1} \right.$$

$$\left. + \frac{T_2^2}{T_1 - T_2} (\varepsilon^{T/T_2} - 1) \varepsilon^{-t/T_2} \right]_T^\infty$$

$$= \frac{V_0 T_1}{T} \left[\frac{T_1^2}{T_1 - T_2} (\varepsilon^{T/T_1} - 1) \varepsilon^{-T/T_1} \right.$$

$$\left. - \frac{T_2^2}{T_1 - T_2} (\varepsilon^{T/T_2} - 1)\varepsilon^{-T/T_2} \right]$$

$$= \frac{V_0 T_1}{T} \left[T_1 + T_2 - \frac{T_1^2}{T_1 - T_2} \varepsilon^{-T/T_1} + \frac{T_2^2}{T_1 - T_2} \varepsilon^{-T/T_2} \right]$$

Hence

Area under pulse
$$= \int_0^T V_{(0-T)} + \int_T^\infty V_{(T-\infty)}$$

$$= \frac{V_0 T_1}{T} \left[T \right]$$

$$= V_0 T_1$$

APPENDIX C

Evaluation of $\displaystyle\int_0^\infty \frac{\omega^2 T_1^2 \, d\omega}{2\pi(1 + \omega^2 T_1^2)(1 + \omega^2 T_2^2)}$

$$= \frac{T_1^2}{2\pi(T_1^2 - T_2^2)} \int_0^\infty \left[\frac{1}{1 + \omega^2 T_2^2} - \frac{1}{1 + \omega^2 T_1^2} \right] d\omega$$

$$= \frac{T_1^2}{2\pi(T_1^2 - T_2^2)} \left[\frac{1}{T_2} \tan^{-1} \omega T_2 - \frac{1}{T_1} \tan^{-1} \omega T_1 \right]_0^\infty$$

$$= \frac{T_1^2}{2\pi(T_1^2 - T_2^2)} \left[\frac{\pi}{2T_2} - \frac{\pi}{2T_1} \right]$$

$$= \frac{T_1}{4T_2(T_1 + T_2)}$$

APPENDIX D

Evaluation of $\displaystyle\int_0^\infty \frac{T_1^2\, d\omega}{2\pi(1 + \omega^2 T_1^2)(1 + \omega^2 T_2^2)}$

$$= \frac{T_1^2}{2\pi(T_1^2 - T_2^2)} \int_0^\infty \left[\frac{T_1^2}{1 + \omega^2 T_1^2} - \frac{T_2^2}{1 + \omega^2 T_2^2}\right] d\omega$$

$$= \frac{T_1^2}{2\pi(T_1^2 - T_2^2)} \left[T_1 \tan^{-1} \omega T_1 - T_2 \tan^{-1} \omega T_2\right]_0^\infty$$

$$= \frac{T_1^2}{2\pi(T_1^2 - T_2^2)} \left[\frac{\pi T_1}{2} - \frac{\pi T_2}{2}\right]$$

$$= \frac{T_1^2}{4(T_1 + T_2)}$$

APPENDIX E

Evaluation of $\displaystyle\int_0^\infty \frac{\omega T_1^2\, d\omega}{(1 + \omega^2 T_1^2)(1 + \omega^2 T_2^2)}$

$$= \frac{T_1^2}{T_1^2 - T_2^2} \int_0^\infty \left[\frac{\omega T_1^2}{1 + \omega^2 T_1^2} - \frac{\omega T_2^2}{1 + \omega^2 T_2^2}\right] d\omega$$

$$= \frac{T_1^2}{T_1^2 - T_2^2} \int_0^\infty \left[\frac{\omega}{\omega^2 + \dfrac{1}{T_1^2}} - \frac{\omega}{\omega^2 + \dfrac{1}{T_2^2}}\right] d\omega$$

$$= \frac{T_1^2}{T_1^2 - T_2^2} \left[\frac{1}{2}\log_\varepsilon\left(\omega^2 + \frac{1}{T_1^2}\right) - \frac{1}{2}\log_\varepsilon\left(\omega^2 + \frac{1}{T_2^2}\right)\right]_0^\infty$$

$$= \frac{T_1^2}{T_1^2 - T_2^2} \left[-\frac{1}{2}\log_\varepsilon\left(\frac{1}{T_1^2}\right) + \frac{1}{2}\log_\varepsilon\left(\frac{1}{T_2^2}\right)\right]$$

$$= \frac{T_1^2}{T_1^2 - T_2^2}\left[\log_\varepsilon \frac{T_1}{T_2}\right]$$

APPENDIX F

Evaluation of $\left[\dfrac{10^{-13} T_1^2}{T_1^2 - T_2^2} \log_\varepsilon \dfrac{T_1}{T_2}\right]^{1/2}$ when $T_1 = T_2$

Consider

$$\frac{T_1^2}{T_1^2 - T_2^2} \log_\varepsilon \frac{T_1}{T_2} = \frac{T_1^2}{T_1 + T_2}\left[\frac{\log_\varepsilon T_1 - \log_\varepsilon T_2}{T_1 - T_2}\right]$$

Now let T_1 tend towards T_2, then $\dfrac{\log_\varepsilon T_1 - \log_\varepsilon T_2}{T_1 - T_2}$ clearly tends towards the differential of $\log_\varepsilon T_1$ with respect to T_1.

$$\therefore \frac{T_1^2}{T_1^2 - T_2^2} \log_\varepsilon \frac{T_1}{T_2} = \frac{T_1^2}{2T_1} \frac{d}{dT_1} (\log_\varepsilon T_1) \text{ when } T_1 = T_2$$

$$= \frac{1}{2}$$

$$\therefore \left[\frac{10^{-13} T_1^2}{T_1^2 - T_2^2} \log_\varepsilon \frac{T_1}{T_2}\right]^{1/2} = \left[\frac{10^{-13}}{2}\right]^{1/2}$$

APPENDIX G

Analysis of cascode circuit in fig. 27 of text.

It is assumed that g_m μ and ρ are the same for both valves. If v_{tri} represents the theoretical shot noise voltage on the grid of the bottom valve and v_a the noise actually appearing at the anode of the bottom valve, the equation for the noise current i_1 through the valve is

$$i_1 = \left(v_{\text{tri}} - \frac{v_a C_a}{\Sigma C}\right) g_m - \frac{v_a}{\rho} \qquad (a)$$

The same current flows through the top valve so a second equation for i_1 is

$$i_1 = v_a g_m - \frac{i_1 R_a - v_a}{\rho} \qquad (b)$$

Solving equation (b) for v_a gives

$$v_a = \frac{i_1\left(1 + \dfrac{R_a}{\rho}\right)}{g_m + \dfrac{1}{\rho}}$$

Substituting in equation (a) gives

$$i_1 = \left[v_{\text{tri}} - \frac{i_1 C_a}{\Sigma C} \frac{1 + \dfrac{R_a}{\rho}}{g_m + \dfrac{1}{\rho}}\right] g_m - \frac{i_1\left(1 + \dfrac{R_a}{\rho}\right)}{\rho\left(g_m + \dfrac{1}{\rho}\right)}$$

$$\therefore i_1 \left[1 + \frac{1 + \dfrac{R_a}{\rho}}{1 + \mu} + \frac{C_a}{\Sigma C} \frac{1 + \dfrac{R_a}{\rho}}{1 + \dfrac{1}{\mu}}\right] = v_{\text{tri}}\, g_m$$

If μ is assumed to be large this gives

$$i_1 = \frac{v_{\text{tri}}\, g_m}{1 + \dfrac{C_a}{\Sigma C}\left(1 + \dfrac{R_a}{\rho}\right)}$$

The output noise voltage is given by

$$i_1 R_a = \frac{v_{\text{tri}}\, g_m\, R_a}{1 + \dfrac{C_a}{\Sigma C}\left(1 + \dfrac{R_a}{\rho}\right)}$$

and hence the effective gain of the cascode is

$$\text{Gain}_1 = \frac{g_m\, R_a}{1 + \dfrac{C_a}{\Sigma C}\left(1 + \dfrac{R_a}{\rho}\right)} \tag{c}$$

Consider now the gain of the circuit to a noise voltage v_{tri} on the grid of the top valve. The impedance Z in the cathode of the top valve is ρ in shunt with $\dfrac{\Sigma C}{g_m C_a}$ due to the bottom valve.

Thus
$$Z = \frac{\rho \Sigma C}{\Sigma C + \mu C_a}$$

APPENDIX G

The noise current caused to flow in the circuit by the top valve is given by

$$i_2 = (v_{\text{tri}} - i_2 Z)g_m - \frac{i_2 R_a + i_2 Z}{\rho}$$

$$\therefore i_2 \left[1 + g_m Z + \frac{R_a + Z}{\rho} \right] = v_{\text{tri}} g_m$$

Hence the output noise voltage is

$$i_2 R_a = \frac{v_{\text{tri}} g_m R_a}{1 + g_m Z + \frac{R_a + Z}{\rho}}$$

Substituting for Z gives

$$i_2 R_a = \frac{v_{\text{tri}} g_m R_a}{1 + \frac{\mu \Sigma C}{\Sigma C + \mu C_a} + \frac{R_a}{\rho} + \frac{\Sigma C}{\Sigma C + \mu C_a}}$$

If now μ is assumed to be large and also μC_a much greater than ΣC, as is the case for low chamber capacity this reduces to

$$i_2 R_a = \frac{v_{\text{tri}} g_m R_a}{1 + \frac{R_a}{\rho} + \frac{\Sigma C}{C_a}}$$

and thus the effective gain of the top valve is

$$\text{Gain}_2 = \frac{g_m R_a}{1 + \frac{R_a}{\rho} + \frac{\Sigma C}{C_a}} \qquad (d)$$

Hence the ratio of the gains is

$$\frac{\text{Gain}_1}{\text{Gain}_2} = \frac{\Sigma C}{C_a} \qquad (e)$$

APPENDIX H

Signal to shot noise ratio of n valves in parallel.

All valves are assumed to have the same g_m μ and ρ. As far as the signal is concerned the n valves can be considered as equivalent to a single valve having a mutual conductance ng_m, anode impedance $\dfrac{\rho}{n}$, amplification factor μ and grid to cathode and anode to grid capacities of nC_g and nC_a respectively.

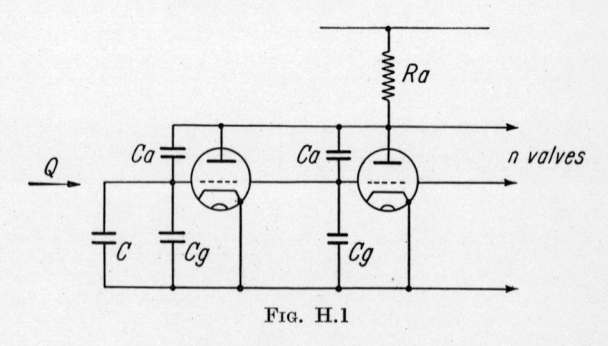

Fig. H.1

The gain M to a voltage appearing between grid and cathode is

$$M = \frac{ng_m R_a \dfrac{\rho}{n}}{R_a + \dfrac{\rho}{n}}$$

$$= \frac{g_m R_a \rho}{R_a + \dfrac{\rho}{n}}. \quad (a)$$

The signal appearing on the grid is $\dfrac{Q}{C + nC_g + (1 + M)nC_a}$ and therefore the anode signal is given by

$$\text{Anode signal} = \frac{QM}{C + nC_g + (1 + M)nC_a} \quad (b)$$

For the noise, consider initially the effect of a single valve having

APPENDIX H

a theoretical shot noise voltage v_{tri} on the grid. This corresponds to a theoretical noise current i_{tri} where $i_{tri} = \dfrac{v_{tri}}{g_m}$.

If i_1 is the noise current which actually flows in the circuit the equation is

$$i_1 = i_{tri} - \frac{i_1 R_a n}{\rho} - \frac{i_1 R_a n C_a n g_m}{C + nC_g + nC_a}$$

$$\therefore i_1 \left[1 + \frac{nR_a}{\rho} + \frac{n^2 R_a C_a g_m}{C + nC_g + nC_a} \right] = i_{tri}.$$

$$\therefore i_1 = \frac{i_{tri}}{1 + \dfrac{nR_a}{\rho} + \dfrac{n^2 R_a C_a g_m}{C + nC_g + nC_a}}$$

$$= \frac{i_{tri}(C + nC_g + nC_a)\dfrac{\rho}{n}}{\left(R_a + \dfrac{\rho}{n}\right)(C + nC_g + nC_a) + nR_a C_a g_m \rho}$$

$$= \frac{i_{tri}(C + nC_g + nC_a)\dfrac{M}{g_m R_a n}}{C + nC_g + nC_a + MnC_a}$$

Thus the noise at the output due to a single valve is

$$i_1 R_a = \frac{\dfrac{i_{tri}}{g_m}(C + nC_g + nC_a) M}{n(C + nC_g + nC_a + MnC_a)}$$

$$= \frac{v_{tri}(C + nC_g + nC_a)M}{n[C + nC_g + nC_a(1 + M)]} \quad (c)$$

The total r.m.s. noise from n valves is $[n]^{1/2}$ times that from a single valve.

$$\therefore \text{Total noise} = \frac{v_{tri} M(C + nC_g + nC_a)}{[n]^{1/2}[C + nC_g + nC_a(1 + M)]} \quad (d)$$

Hence the signal to noise ratio is given by

$$\frac{S}{N}_{(Shot)} = \frac{Q[n]^{1/2}}{v_{tri}(C + nC_g + nC_a)} \quad (e)$$

To find the maximum for equation (e) differentiate with respect to n and equate to zero. This gives

$$\frac{C + nC_g + nC_a}{2[n]^{1/2}} = [n]^{1/2}(C_g + C_a)$$

and hence

$$n = \frac{C}{C_g + C_a} \quad (f)$$

APPENDIX I

Evaluation of

$$V_{(T-\infty)} = \frac{V_0 T_1^2}{T(T_1 - T_2)}\left[\varepsilon^{T/T_1} - 1\right]\varepsilon^{-t/T_1}$$

$$- \frac{V_0 T_1 T_2}{T(T_1 - T_2)}\left[\varepsilon^{T/T_2} - 1\right]\varepsilon^{-t/T_2} \quad (2)$$

when $T_1 = T_2$

$$V_{(T-\infty)} = \frac{V_0 T_1}{T}\left[\frac{d}{dT_1}\left\{T_1(\varepsilon^{T/T_1} - 1)\varepsilon^{-t/T_1}\right\}\right]$$

$$= \frac{V_0 T_1}{T}\left[\left\{\varepsilon^{\frac{T-t}{T_1}} + \frac{T(T-t)\varepsilon^{\frac{T-t}{T_1}}}{-T_1^2}\right\} - \left\{\varepsilon^{-t/T_1} + \frac{T(-t)\varepsilon^{-t/T_1}}{-T_1^2}\right\}\right]$$

$$= \frac{V_0 T_1}{T}\left[\varepsilon^{-t/T_1}(\varepsilon^{T/T_1} - 1)\right.$$

$$\left. + \varepsilon^{-t/T_1}\left(-\frac{t}{T_1} + \frac{t}{T_1}\varepsilon^{T/T_1} - \frac{T}{T_1}\varepsilon^{T/T_1}\right)\right]$$

$$= \frac{V_0 T_1}{T}\varepsilon^{-t/T_1}\left[\varepsilon^{T/T_1} - 1 - \frac{T}{T_1}\varepsilon^{T/T_1} + \frac{t}{T_1}(\varepsilon^{T/T_1} - 1)\right] \quad (36)$$

APPENDIX J

Evaluation of

$$t_m = \frac{T_1 T_2}{T_1 - T_2} \log_\varepsilon \frac{\varepsilon^{T/T_2} - 1}{\varepsilon^{T/T_1} - 1}$$

when $T_1 = T_2$

$$t_m = -T_1^2 \left[\frac{d}{dT_1} \log_\varepsilon (\varepsilon^{T/T_1} - 1) \right]$$

$$= -T_1^2 \left[\frac{T \varepsilon^{T/T_1}}{(\varepsilon^{T/T_1} - 1)(-T_1^2)} \right]$$

$$= \frac{T \varepsilon^{T/T_1}}{\varepsilon^{T/T_1} - 1} \qquad (37)$$

APPENDIX J

Evaluation of

$$V_{(T-\infty)} = \frac{V_0 T_1}{T} \varepsilon^{-t/T_1} \left[\varepsilon^{T/T_1} - 1 - \frac{T}{T_1} \varepsilon^{T/T_1} + \frac{t}{T_1} (\varepsilon^{T/T_1} - 1) \right] \quad (36)$$

when T tends to zero.

$$V_{(T-\infty)} = \frac{V_0 T_1}{T} \varepsilon^{-t/T_1} \left[\frac{T}{T_1} - \frac{T}{T_1}\left(1 + \frac{T}{T_1}\right) + \frac{t}{T_1} \frac{T}{T_1} \right]$$

$$= \frac{V_0 T_1}{T} \varepsilon^{-t/T_1} \left[\frac{tT}{T_1^2} - \frac{T^2}{T_1^2} \right]$$

$$= V_0 \varepsilon^{-t/T_1} \left[\frac{t}{T_1} - \frac{T}{T_1} \right]$$

$$= \frac{V_0 t}{T_1} \varepsilon^{-t/T_1} \qquad (38)$$

APPENDIX K

Suppose the initial operating conditions of the valve are g_{m1} and I_{g1}.

Shot noise $= \dfrac{Q_1}{[g_{m1}]^{1/2}}$ where Q_1 is a constant and grid current noise $= P_1[I_{g1}]^{1/2}$ where P_1 is also a constant. Since the noise components are equal,

$$P_1[I_{g1}]^{1/2} = \frac{Q_1}{[g_{m1}]^{1/2}}$$

$$\therefore \frac{Q_1}{P_1} = [I_{g1} g_{m1}]^{1/2} \tag{a}$$

and the total noise is $= P_1[2I_{g1}]^{1/2}$ \qquad (b)

If now the frequency band is lowered in the spectrum by changing T_1 to $n_1^2 T_1$ this gives

$$\text{Shot noise} = \frac{Q_1}{n_1[g_{m1}]^{1/2}}$$

and \qquad grid current noise $= n_1 P_1 [I_{g1}]^{1/2}$

Finally I_{g1} and g_{m1} are changed to I_{g2} and g_{m2} and this again makes the noise components equal.

$$\therefore n_1 P_1 [I_{g2}]^{1/2} = \frac{Q_1}{n_1 [g_{m2}]^{1/2}}$$

$$\therefore n_1^2 = \frac{Q_1}{P_1} \frac{1}{[I_{g2} g_{m2}]^{1/2}}$$

$$= \left[\frac{I_{g1}}{I_{g2}} \frac{g_{m1}}{g_{m2}} \right]^{1/2} \tag{c}$$

and the total noise is $= n_1 P_1 [2I_{g2}]^{1/2}$ \qquad (d)

Hence
$$\frac{\text{Initial Noise}}{\text{New Noise}} = \frac{1}{n_1}\left[\frac{I_{g1}}{I_{g2}}\right]^{1/2}$$

$$= \left[\frac{I_{g1}}{I_{g2}}\right]^{1/2}\left[\frac{I_{g2}\,g_{m2}}{I_{g1}\,g_{m1}}\right]^{1/4}$$

$$= \left[\frac{I_{g1}\,g_{m2}}{I_{g2}\,g_{m1}}\right]^{1/4}$$

$$= \frac{\left[\dfrac{I_{g1}}{g_{m1}}\right]^{1/4}}{\left[\dfrac{I_{g2}}{g_{m2}}\right]^{1/4}}$$

$$= [\phi]^{1/4} \tag{e}$$

where
$$\phi = \frac{\dfrac{I_{g1}}{g_{m1}}}{\dfrac{I_{g2}}{g_{m2}}}$$

APPENDIX L

Transient analysis of the circuit in fig. 34 of text.

Let I be the change in anode current produced by the step voltage on the grid of the valve and let i_1 and i_2 be the current changes through the resistor R_d and the inductance L_d respectively at time t after the application of the step function.

The following equations apply

$$i_1 R_d = \frac{1}{C_d}\int\left[I - (i_1 + i_2)\right]dt \tag{a}$$

and
$$i_1 R_d = L_d \frac{di_2}{dt} \tag{b}$$

From equation (a)

$$\frac{di_1}{dt} R_d = \frac{I - i_1 - i_2}{C_d}$$

hence
$$i_2 = I - i_1 - C_d R_d \frac{di_1}{dt}$$

Substituting this value for i_2 in equation (b) gives

$$i_1 R_d = L_d \left[-\frac{di_1}{dt} - C_d R_d \frac{d^2 i_1}{dt^2} \right]$$

$$\therefore \frac{d^2 i_1}{dt^2} + \frac{1}{C_d R_d} \frac{di_1}{dt} + \frac{1}{L_d C_d} = 0 \qquad (c)$$

From equation (c) the condition for critical damping is

$$\frac{1}{L_d C_d} = \left[\frac{1}{2 C_d R_d} \right]^2$$

giving
$$R_d = \frac{1}{2} \left[\frac{L_d}{C_d} \right]^{1/2} \qquad (d)$$

When critically damped the solution of equation (c) is aperiodic and has the form $i_1 = (At + B)\, \varepsilon^{-\frac{t}{2C_d R_d}}$ where A and B are constants of integration.

When $\quad t = 0 \quad i_1 = 0 \quad$ and hence $\quad B = 0$.

Also when $\quad t = 0 \quad \dfrac{di_1}{dt} = \dfrac{I}{C_d R_d}$

Differentiating gives $\quad \dfrac{di_1}{dt} = A \left[\varepsilon^{-\frac{t}{2C_d R_d}} - \dfrac{t}{2 C_d R_d} \varepsilon^{-\frac{t}{2C_d R_d}} \right]$

$$\therefore \frac{I}{C_d R_d} = A.$$

Hence the solution for i_1 is

$$i_1 = \frac{I}{C_d R_d} t\, \varepsilon^{-\frac{t}{2C_d R_d}}$$

and the voltage pulse on the anode is

$$i_1 R_d = \frac{I}{C_d} t\, \varepsilon^{-\frac{t}{2C_d R_d}} \qquad (e)$$

APPENDIX M

Impedance of a shorted line shunted by a resistor equal to the characteristic impedance of the line.

From any text book on transmission line theory the input impedance of a shorted line at high frequencies is given by

$$Z_{in} = jZ_0 \tan \omega l \, [L'C']^{1/2}$$

where l is the length of the line, L' and C' are the inductance and capacitance per unit length, and Z_0 is the characteristic impedance.

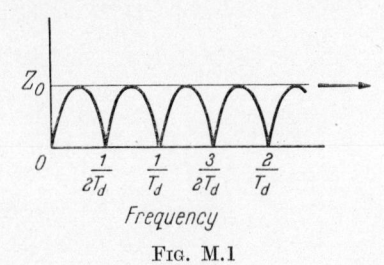

Fig. M.1

If T_d is the time delay from one end of the line to the other then $T_d = l[L'C']^{1/2}$.

Hence

$$Z_{in} = jZ_0 \tan \omega T_d$$

The impedance of Z_{in} shunted by a resistor Z_0 is

$$Z = \frac{jZ_0^2 \tan \omega T_d}{Z_0 + jZ_0 \tan \omega T_d}$$

$$= \frac{jZ_0 \tan \omega T_d (1 - j \tan \omega T_d)}{1 + \tan^2 \omega T_d}$$

$$= Z_0 \sin \omega T_d \cos \omega T_d \, (\tan \omega T_d + j)$$

$$= Z_0 \sin \omega T_d \, (\sin \omega T_d + j \cos \omega T_d)$$

$(\sin \omega T_d + j \cos \omega T_d)$ is numerically always equal to unity but changes its sign from positive to negative every half cycle of ωT_d. The impedance Z therefore varies with frequency as a series of half sine waves as shown in fig. M.1.

Consequently the square of the impedance is simply given by

$$Z^2 = (Z_0 \sin \omega T_d)^2$$

APPENDIX N

Evaluation of $\int_0^\infty \dfrac{\sin^2 \omega T_d \, d\omega}{2\pi(1 + \omega^2 T_2^2)}$

$$= \frac{1}{4\pi} \int_0^\infty \frac{1 - \cos 2\omega T_d}{1 + \omega^2 T_2^2} d\omega$$

$$= \frac{1}{4\pi} \left[\frac{1}{T_2} \tan^{-1} \omega T_2 \right]_0^\infty - \frac{1}{4\pi} \int_0^\infty \frac{\cos 2\omega T_d \, d\omega}{1 + \omega^2 T_2^2}$$

$$= \frac{1}{8T_2} - \frac{1}{4\pi} \int_0^\infty \frac{\cos 2\omega T_d \, d\omega}{1 + \omega^2 T_2^2}$$

Let $\omega T_2 = x$ and $\dfrac{2T_d}{T_2} = a$ then $d\omega = \dfrac{dx}{T_2}$

Consider $\int_0^\infty \dfrac{\cos 2\omega T_d \, d\omega}{1 + \omega^2 T_2^2}$

$$= \frac{1}{T_2} \int_0^\infty \frac{\cos ax \, dx}{1 + x^2}$$

$$= \frac{1}{T_2} \left[\frac{\pi}{2} \varepsilon^{-a} \right]$$

$$= \frac{\pi}{2T_2} \varepsilon^{-\frac{2T_d}{T_2}}$$

since $\int_0^\infty \dfrac{\cos ax \, dx}{1 + x^2}$ is a definite integral which is tabulated in most text books on calculus.

Hence the initial integral is

$$= \frac{1}{8T_2} \left[1 - \varepsilon^{-\frac{2T_d}{T_2}} \right]$$

APPENDIX O

Composite curve to give a point of inflexion.

Suppose the sum of the two separate Gaussian contours is given by

$$\varepsilon^{-\frac{q^2}{2}} + U\varepsilon^{-\frac{(q-w)^2}{2}}$$

where $\qquad U =$ Relative height of the peaks

$$q = \frac{x}{v} = \frac{x}{\text{r.m.s. value}}$$

and $\qquad w = \dfrac{\text{Peak separation}}{\text{r.m.s. value}}.$

For a point of inflexion the first and second differentials must be zero.

$$\therefore q\varepsilon^{-\frac{q^2}{2}} + U(q-w)\varepsilon^{-\frac{(q-w)^2}{2}} = 0. \tag{a}$$

Also $\quad \varepsilon^{-\frac{q^2}{2}} - q^2\varepsilon^{-\frac{q^2}{2}} + U\left[\varepsilon^{-\frac{(q-w)^2}{2}} - (q-w)^2\varepsilon^{-\frac{(q-w)^2}{2}}\right] = 0$

giving $\quad (1-q^2)\varepsilon^{-\frac{q^2}{2}} + U\left[1-(q-w)^2\right]\varepsilon^{-\frac{(q-w)^2}{2}} = 0. \tag{b}$

From equations (a) and (b)

$$\frac{q}{1-q^2} = \frac{(q-w)}{1-(q-w)^2}$$

$$\therefore q - q^3 - qw^2 + 2q^2w = q - q^3 - w - q^2w$$

$$\therefore q^2 - qw + 1 = 0$$

Solving this gives

$$q = \frac{w \pm (w^2 - 4)^{1/2}}{2}$$

and $\qquad q - w = \dfrac{-w \pm (w^2 - 4)^{1/2}}{2}.$

150 SIGNAL AND NOISE IN NUCLEAR COUNTER AMPLIFIERS

Substituting in equation (a) gives

$$U = \frac{-q}{q-w} \varepsilon^{-\frac{q^2}{2} + \frac{(q-w)^2}{2}}$$

$$= -\left[\frac{w \pm (w^2-4)^{1/2}}{-w \pm (w^2-4)^{1/2}}\right] \varepsilon^{\mp \frac{w(w^2-4)^{1/2}}{2}}$$

This equation is used to compute the points for the curve in fig. 40.

APPENDIX P

Evaluation of $\int_0^\infty t^2 \varepsilon^{-\frac{2t}{T_1}} dt$

$$= \left[-\frac{T_1 t^2}{2} \varepsilon^{-\frac{2t}{T_1}}\right]_0^\infty - \int_0^\infty -T_1 t \, \varepsilon^{-\frac{2t}{T_1}} dt$$

$$= 0 + T_1 \int_0^\infty t \, \varepsilon^{-\frac{2t}{T_1}} dt$$

$$= T_1 \left[-\frac{T_1 t}{2} \varepsilon^{-\frac{2t}{T_1}}\right]_0^\infty - T_1 \int_0^\infty -\frac{T_1}{2} \varepsilon^{-\frac{2t}{T_1}} dt$$

$$= 0 + \frac{T_1^2}{2} \int_0^\infty \varepsilon^{-\frac{2t}{T_1}}$$

$$= \frac{T_1^2}{2} \left[-\frac{T_1}{2} \varepsilon^{-\frac{2t}{T_1}}\right]_0^\infty$$

$$= \frac{T_1^3}{4}$$

APPENDIX Q

Evaluation of $\int_0^\infty \varepsilon^{-\frac{2t}{T_1}} \left(1 - \frac{t}{T_1}\right)^2 dt$

$$= \int_0^\infty \varepsilon^{-\frac{2t}{T_1}} - \frac{2}{T_1} \int_0^\infty t\varepsilon^{-\frac{2t}{T_1}} + \frac{1}{T_1^2} \int_0^\infty t^2 \varepsilon^{-\frac{2t}{T_1}}$$

All the above three integrals have already been evaluated in Appendix P. Thus

$$I = \frac{T_1}{2} - \frac{T_1}{2} + \frac{T_1}{4}$$

$$= \frac{T_1}{4}$$

APPENDIX R

Evaluation of $\int_0^\tau \left[\frac{d}{dt}\left(\frac{t}{\tau}\right)\right]^2 dt + \int_\tau^\infty \left[\frac{d}{dt}\left\{\varepsilon^{-\frac{(t-\tau)}{T_1}} \left(1 - \frac{t-\tau}{T_1}\right)\right\}\right]^2 dt$

Taking each in turn

$$\int_0^\tau \left[\frac{d}{dt}\left(\frac{t}{\tau}\right)\right]^2 dt = \int_0^\tau \frac{1}{\tau^2} dt$$

$$= \left[\frac{t}{\tau^2}\right]_0^\tau$$

$$= \frac{1}{\tau}$$

Also

$$\int_\tau^\infty \left[\frac{d}{dt}\left\{\varepsilon^{-\frac{(t-\tau)}{T_1}}\left(1-\frac{t-\tau}{T_1}\right)\right\}\right]^2 dt$$

$$= \int_0^\infty \left[\frac{d}{dt}\left\{\varepsilon^{-\frac{t}{T_1}}\left(1-\frac{t}{T_1}\right)\right\}\right]^2 dt$$

$$= \int_0^\infty \left[-\frac{1}{T_1}\varepsilon^{-\frac{t}{T_1}}\left(1-\frac{t}{T_1}\right)-\frac{1}{T_1}\varepsilon^{-\frac{t}{T_1}}\right]^2 dt$$

$$= \int_0^\infty \varepsilon^{-\frac{2t}{T_1}}\left(\frac{t}{T_1^2}-\frac{2}{T_1}\right)^2 dt$$

$$= \frac{1}{T_1^4}\int_0^\infty t^2 \varepsilon^{-\frac{2t}{T_1}} dt - \frac{4}{T_1^3}\int_0^\infty t\varepsilon^{-\frac{2t}{T_1}} dt + \frac{4}{T_1^2}\int_0^\infty \varepsilon^{-\frac{2t}{T_1}} dt$$

The three integrals have already been evaluated in Appendix P. Thus the second composite integral is

$$= \frac{1}{4T_1} - \frac{1}{T_1} + \frac{2}{T_1}$$

$$= \frac{5}{4T_1}$$

and the original integral is $\dfrac{1}{\tau} + \dfrac{5}{4T_1}$

INDEX

Ageing of valves 34
α-particle 2, 82, 109, 110, 111, 118
Amplifier
— bandwidth 3, 14, 20, 125
— equivalent circuit 14
— function 2, 6
— gain 3, 14, 120
— response to step input 66
— unavoidable integration 94

Background, counting in presence of 10, 111–119
β-particle 2, 111
BOLTZMANN's constant 4, 21

CAMPBELL's theorem 87
Cascode 57, 58
Circuit noise, see Noise
Charged particle 1, 2
Collection time 5, 8, 24, 114
— — effect of variation 69, 70
— — electron chamber 11
— — ion chamber 10
— — proportional counter 108–111
Collector plate 5

Delay line pulse shaping 7, 72–78
— — practical application 77, 78
Differentiating circuit pulse shaping 7–9
Differentiating time constant 8, 9, 23, 24
— — — effect on pulse from electron chamber 12, 14–18
— — — effect on pulse from ion chamber 10, 14–18
— — — effect on pulse from proportional counter 107, 108–111
— — — effect on pulse from scintillation counter 122
Discriminator 16, 17, 49, 51, 83, 87, 94, 97, 103
— bias curve 51, 52, 102, 115
Drift velocity
— — electrons 11
— — negative ions 5
— — positive ions 5

Equivalent time constant 114
Electron 2, 4, 10, 21, 24, 29, 44, 105
— secondary 105

Electron chamber 10–14
— — gas filling 10
— — gridded 12, 13
— — operation 10–13
— — theoretical pulse 14, 15
— — voltage pulse 11–13
Electronic charge 5, 24, 81
Energy discrimination 2, 79

Fission counter 118, 119
— fragments 2, 111
— pulses 111, 118
FRISCH grid 12

γ-ray 2, 111
Gaussian distribution 50, 98, 101, 112, 116, 118
GEIGER-MÜLLER counter 1
Grid current 29–31
— — components 29–31
— — reduction 33
— — EF37 34
— — ME1400 35, 36
— — EC91 37
— — 6AK5 39
Grid emission 29, 30, 33, 34, 69

Integrating time constant 14
— — — effect on pulse from ionisation chamber 14–18
Ion chamber 4
— — operation of air filled 4, 5, 10
— — theoretical pulse 10, 15
— — voltage pulse 5, 9, 10
Ionisation 1, 4, 10, 13
— primary 1, 107
— secondary 1, 2, 105, 106
— valve 29, 33
Ionisation chamber 1, 2, 4–10
Ion pair 1, 5, 81, 82, 83–85, 102, 104
— — energy 1, 4, 81
Isotope 82

Kicksorter 3, 49, 50, 79

Line spectrum 49, 50
— — effect of noise 50, 51

Negative ion 4
Neutron
— fast 2
— slow 2

154   SIGNAL AND NOISE IN NUCLEAR COUNTER AMPLIFIERS

Noise   2, 9, 19–48, 50
— basic phenomena   19, 20, 50, 87–90
— calculation   20, 21, 22, 26–28, 31, 32, 38, 41, 42
— counting of crests   51, 52, 86, 87, 90–96
— experimental figures   43, 44
— flicker   40–44
— grid current   29–40, 87, 88
— induced grid   44–46
— measurement   20
— partition   26, 27
— photomultiplier   123, 124
— resistance   26
— semi-conductor   47, 48
— shot   24–29, 46, 47
— space charge smoothing   25, 27, 46, 88
— spectral density   20, 21, 22–24, 31, 41, 45, 47, 48
— thermal   21–24, 46, 47
Nuclear counter   1, 2, 4
— — geometry   52, 115, 117

Phosphor   1, 120
Photoelectron   1, 120
Photomultiplier   1, 2, 120
— gain   120
Photon   1, 120
— energy   1
Plateau   52, 102, 109, 113, 115, 118
Poisson's formula   6, 116, 117
Positive ion   4, 29, 106
Proportional counter   1, 2, 105–119
— — filling gas   106
— — gas gain   105, 106
— — geometry   105, 115
— — operation   105, 106
— — voltage pulse   106–108
Proton   2, 111

Random distribution   6, 9
Ratemeter   2, 17, 97, 103, 109
Register   2, 17, 97
Resolving power   7, 8, 17, 67, 69
— — electron chamber   12, 13
— — ion chamber   7
— — proportional counter   108
— — scintillation counter   122
Resolving time   16–18, 20, 67, 114, 116
— — bandwidth relationship   59–66
— — delay line pulse shaping   72, 76, 77
— — ringing coil pulse shaping   71, 72

Response time   94, 95, 112, 113, 114, 116, 124, 125
Ringing coil pulse shaping   7, 71, 72

Saturation potential   4
Scaler   2, 17, 97
Scintillation counter   1, 2, 120–123
— — operation   120, 121
— — voltage pulse   121
Scintillation time   120–122
Secondary emission   120, 124
Selection of valves for low grid current   34
Sensitivity   20, 79–104
— calculation   81, 82, 101, 102
— energy measurements   79–86
— experimental derivation   97–101
— experimental figures for EC91, 6AK5 and ME1400   84, 85, 103, 104
— measurement   82, 83
— separation of energy groups   80
— total activity measurements   86–104
Signal to noise ratio   2, 7, 15, 23, 49–78
— — — — best obtainable   68–69
— — — — calculation   56
— — — — cascode   57, 58
— — — — choice of input valve   67
— — — — delay line pulse shaping   72–77
— — — — effect of feedback   55
— — — — energy measurements   49–51
— — — — function of bandwidth   59–64
— — — — function of collection time   69, 70
— — — — function of resolving power   68, 85
— — — — pentode   53
— — — — ringing coil pulse shaping   71, 72
— — — — scintillation counter   123–126
— — — — total activity measurements   51–52
— — — — triode   54–56
— — — — valves in parallel   59
Space charge limitation   24
Specific ionisation
— α-particle   111
— β-particle   111

Temperature limitation   24
Thermal agitation   4, 21
— — energy   4

Thermal voltmeter 43, 83, 103
THEVININ's theorem 21
Transit time
—— electrons 29, 44, 88, 124
—— positive ions 38

Uranium 82

Valve noise, *see* Noise

X-ray 2